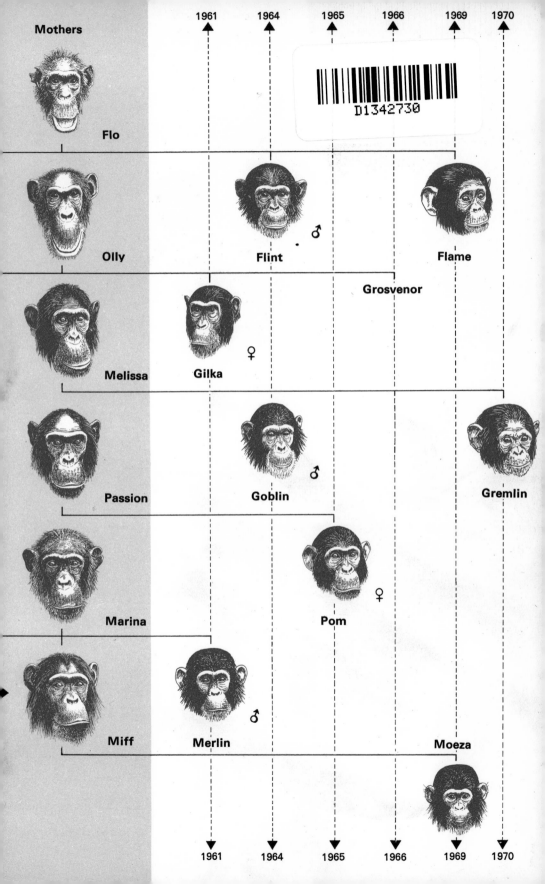

Mothers

1961 1964 1965 1966 1969 1970

Flo

Olly Flint ♂ Flame

Grosvenor

Melissa Gilka ♀

Passion Goblin ♂ Gremlin

Pom ♀

Marina

Miff Merlin ♂ Moeza

1961 1964 1965 1966 1969 1970

# In the Shadow of Man

*by the same author*

THE INNOCENT KILLERS
(with Hugo van Lawick)

GRUB The Bush Baby

# In the Shadow of Man

*Jane van Lawick-Goodall*

Photographs by
Hugo van Lawick

BOOK CLUB ASSOCIATES, *London*

This edition published 1971 by
Book Club Associates
by arrangement with Messrs Wm Collins Sons & Co Ltd

© Hugo and Jane van Lawick-Goodall 1971

ISBN 0 00 211357 0

Set in Monotype Spectrum

Made and Printed in Great Britain by
William Collins Sons & Co Ltd Glasgow

For Hugo, Vanne and Louis and to the
memory of David Greybeard

# Contents

# Acknowledgements

I should have been unable to study chimpanzees or write this book without the help and encouragement of a great many people, and I should like to express, however inadequately, my most profound thanks to all of them. Firstly, of course, my gratitude lies with Dr L. S. B. Leakey. It was he who suggested that I should study chimpanzees in the first place, he who found money to finance my early fieldwork and he who arranged for me to write up my results for a PhD dissertation at Cambridge University. Last – but by no means least! – it was through Louis's recommendation that Hugo came to photograph the chimpanzees.

I am tremendously grateful to the Tanzanian Government, its President, Mwalimu Julius Nyerere, and many of its officials, for permitting us to conduct our research in the Gombe Stream area, and for giving us at all times help and assistance. I was initially helped by the Head and officials of the Tanzania Game Department, and I am particularly grateful to David Anstey who assisted my mother and I when we first set up camp, and to the African Game Scouts Adolf, Saulo David and Marcel who were stationed in the Gombe Stream when the area was a Game Reserve. More recently, since the Gombe became a National Park, we have received complete co-operation and assistance from Dr John Owen, ex-Director of Tanzania National Parks, and his successor, Mr S. ole Saibul. My thanks are also due to other members of the Tanzania National Parks staff, especially to Mr J. Stevenson, Director of Southern National Parks, and to the African Game Rangers stationed in the Park.

My thanks are due also to Government officials and many friends in Kigoma who, through the years, have done so much to help our research as well as to give us personal assistance in countless ways.

I am very much indebted to Mr Leighton Wilkie who granted

the funds which enabled me to venture forth on my fieldwork in 1960 and who has, more recently, granted a further donation to the Gombe Stream Research Centre. My gratitude to the National Geographic Society is sincere and of the greatest magnitude. The Society took over the financing of my research in 1961, supported the entire project at Gombe until 1968, and is continuing to make a major annual contribution to-day. In particular I should like to express my deepest thanks to Dr Melville Bell Grosvenor, President until 1967, to Dr Melvin Payne, his successor, and to Dr Leonard Carmichael, Chairman of the Committee for Research and Exploration for their long-standing encouragement and support. I thank, too, the other members of the Committee and the staff and members of the Society, especially Mr Robert Gilka, Miss Joanne Hess and Miss Mary Griswold, who have gone out of their way to help us on many occasions.

In 1969 we received a substantial grant from the Science Research Council of Great Britain, and more recently we have received further financial support from the Wenner-Gren Foundation, the East African Wildlife Society, the L. S. B. Leakey Foundation, and the World Wildlife Fund. To all of these organisations, as well as to a number of private individuals who have, from time to time, made contributions to our research, I express our deep gratitude.

My profound thanks are also due to Professor Robert Hinde of Cambridge University who not only supervised the analysis and writing up of my results for my doctoral degree, but who has been instrumental in obtaining some funds for maintaining the research and devoted much time and effort to helping us in many other ways. I am deeply grateful also to Professor David Hamburg of Stanford University who has long expressed interest in the research and who has made it possible for the Centre to become affiliated with Stanford University. Professor Hamburg has been of great assistance in obtaining some of the necessary funds for the maintenance of the Gombe Stream Research Centre, and is continuing to help us actively in this way at the present time. Both Professor Hinde and Professor Hamburg have agreed to act as scientific advisers to the Centre. I am very appreciative also of the keen interest in our work and encouragement shown by Professor S. A. Msangi, Dean of the Faculty of Science of the University College of Dar es Salaam.

It is difficult to find adequate words to thank Hugo for all that he has done for me and for the research. Not only has he built up a magnificent collection of photographs and a unique documentary film record of chimpanzee behaviour, but it is to a large extent due to his constant help, administrative ability and persistence that the Gombe Stream Research Centre was formed and is flourishing to-day. I am sure I could never have embarked on such a project on my own. Hugo's patience and understanding both of his chimpanzee subjects and his wife are truly remarkable. I must try, also, to thank my mother for all that she has done for me through the years, most particularly for her courage, patience and cheerfulness through the early months when she shared with me the most primitive of living conditions. On many occasions, too, her advice and suggestions have proved invaluable. I am indebted to both my mother and to Hugo for the many comments and valuable criticisms they made when I was writing this book.

There are many, many people who have, directly or indirectly, made contributions to our research and helped us personally, and it is not possible to mention them all by name. I should, however, like to thank Dr Bernard Verdcourt of the Royal Botanic Gardens at Kew, who initially drove my mother and myself to the Gombe Stream and who identified many food plants, and to Dr Gillet of the East African Herbarium who also identified plant specimens. My thanks are due also to the Pfizer Laboratories who so generously supplied us with complimentary polio vaccine during the terrible epidemic at the Gombe Stream. I am also extremely grateful to Professor Douglas Roy, Drs Anthony and Sue Harthoorn, and Dr Bradly Nelson, all of whom assisted in the tranquillisation and operation on the chimpanzee Gilka.

Next I should like to express my most sincere thanks to all the African staff and helpers who have done so much through the years to make our work easier and our life more pleasant. Particularly my thanks are due to Hassan and Dominic, Rashidi, Soko, Wilbert and Short who, during the early years were, at times, my only companions in the bush. I thank also the many others: Sadiki, Ramadthani, Juma, Mpofu, Hilali, Alphonse, Jumanne, Kasim Ramadhani, Kasim Selemani, Yahaya, Aporual, Habibu and Adreano. I should also like to express our gratitude to Iddi Matata and Mbrisho for the courtesy they have always shown us and the

# Acknowledgements

way in which they have made us feel welcome in their country. I thank too Mucharia and Moro who helped me so much by looking after my son whilst I was working on this book.

My thanks are due to Kris Pirozynski who looked after camp in the early days, and to my sister Judy who took some of the earliest photos of chimpanzees in the wild. I am grateful to Nic and Margaret Pickford who worked at the Centre as administrators, to Baron and Baroness Godert and Bobbie van Lawick-de Marchant et d'Ansembourg who also helped us in this capacity, and to Michael Richmond and Miss Dan for their administrative help in Nairobi.

I should like to acknowledge the work of Dr Peter Marler and Dr Michael Simpson. Both worked at the Gombe Stream on their own studies: Dr Marler recorded chimpanzee calls over a two-month period, and Dr Simpson spent eighteen months working on social grooming. The results of these studies will be of great value to the Gombe Stream Research Centre and will provide us with added information on chimpanzee behaviour. My thanks are due to Tim and Bonnie Ransom, Leanne Taylor and Nic Owens, whose observations on the baboon troops at Gombe enabled us to learn more about the chimpanzee's predatory attempts on baboons and other chimpanzee-baboon interactions. Tim was extremely helpful when we changed the feeding system in 1968. I thank, too, John McKinnon who worked at the Gombe Stream on insect behaviour, but devoted much time to helping us with our study of chimpanzee behaviour.

Finally, I reach the point when I must attempt to express my thanks and gratitude to the students who, as research assistants, have so greatly increased our understanding of chimpanzee behaviour. It is difficult, in a short space, to convey adequately my appreciation of the hard work, patience and dedication that has, in many instances, gone into the careful accumulation of the long-term records on individual chimpanzees on which I have drawn so freely in the writing of this book. Without these students a longitudinal study of this sort could not be undertaken and this book could never have been written. At this point I should especially like to acknowledge the help of Edna Koning, Sonia Ivey, Alice Sorem and Pat McGinnis who worked so hard in the early pioneering days.

Some assistants have remained with us for a short time only, yet they too have made their contributions to the research programme: Sue Chaytor, Sally Avery, Pamela Carson, Patti Moehlman, Nicoletta Maraschin, June Cree, Janet Brooks, Sanno Keeler and Neville Washington. Others have stayed a full year as my research assistants, helping to accumulate data in the all-important long-term records: Caroline Coleman, Cathleen Clarke, Carole Gale, Dawn Starin, Ann Simpson.

Yet other students, after working for a year on the long-term records, have stayed on to start their own research projects on different aspects of behaviour. Alice Sorem worked on mother-infant interactions: I am particularly grateful to her hard work during the difficult time when each chimpanzee was given oral polio vaccine. Geza Teleki has already written a report on meat-eating behaviour in chimpanzees, and he is now working on chimpanzee range at Gombe. Lori Baldwin is studying the relationship between adult female chimpanzees, and David Bygott is studying those between adult males, with emphasis on dominance and aggression. Both Lori and David will work for their doctoral degrees at Cambridge University, under the supervision of Professor Hinde. Another student who is now working for his PhD degree at Cambridge, also under Professor Hinde, is Patrick McGinnis. His subject is reproductive behaviour. I have a special word of thanks for Pat; he has already worked at the Gombe Stream for nearly four years and, on many occasions, not only been in charge of the research, but kept the place running almost single-handed when other students were away or sick. The Gombe Stream seems almost strange when Pat is away at Cambridge.

Recently four new students have joined our team, and they too plan to stay on for a second year to study different aspects of behaviour: Harold Bauer, Ann Pusey, Margaretha Hankey and Richard Wrangham. I should like to thank all of them for their hard work and co-operation. I should also like to express particular appreciation to Dr Helmut Albrecht who has recently joined our research team as Senior Scientist and who has already proved himself an ideal person for this somewhat arduous task. I am convinced that his two-year sojourn at Gombe will be a happy one for all concerned.

Now comes the most difficult part of these acknowledgements –

that of trying to express my debt of gratitude to Ruth Davis, who lost her life whilst studying chimpanzees at the Gombe Stream. Ruth was not strong, yet she was one of the hardest workers we have had, and sometimes drove herself to near exhaustion. She chose to study the individualities of the adult male chimpanzees, she spent long arduous hours in the mountains observing and following her subjects, and sometimes typed out her notes until late in the evenings. It may have been due to physical exhaustion that one day, in 1968, Ruth fell from the edge of a precipice and was instantly killed. Her body was only found after a search of six days, during which many people took part, including the Kigoma police, National Parks officials and numerous volunteers from the surrounding villages. We were immensely grateful to all these people.

It is impossible to find proper words to express my deep regret and sorrow at this terrible tragedy: Ruth's death was a great personal loss to all of us who knew her, particularly to her fiancé, Geza Teleki. She was buried in the National Park in the country that she loved so much during her life: her grave is surrounded by the forest and reverberates, from time to time, with the calling of the chimpanzees as they pass by.

My greatest admiration and deepest sympathy goes to Ruth's parents who visited the Gombe Stream for the first time under the tragic circumstances of their daughter's burial. Despite their distress they were able to assure us that we should not feel ourselves at all responsible for the accident: Ruth, they told us, had known, during her year with the chimpanzees, the happiest time of her life and, in the work, had found true fulfilment.

Ruth met her death whilst doing work that she really wanted to do, in the wilderness of the mountains that she loved: her patient study has given us added insight into the personalities of some of the chimpanzees. Ruth, I know, would wish me to end my acknowledgements with a tribute to the chimpanzees themselves; those amazing creatures who can teach us so much about ourselves even whilst we become increasingly fascinated by them in their own right. To David Greybeard and Flo, in particular, we owe much.

*Chapter 1* Beginnings

Since dawn I had climbed up and down the steep mountain slopes and pushed my way through the dense valley forests. Again and again I had stopped to listen, or to gaze through binoculars at the surrounding countryside. Yet I had neither heard nor seen a single chimpanzee and now it was already five o'clock. In two hours darkness would fall over the rugged terrain of the Gombe Stream Reserve. I settled down at my favourite vantage point, the Peak, hoping that at least I might see a chimpanzee make his nest for the night before I had to stop work for the day.

I was watching a troop of monkeys in the forested valley below when suddenly I heard the screaming of a young chimpanzee. Quickly I scanned the trees with my binoculars, but the sound had died away before I could locate the exact place, and it took several minutes of searching before I saw four chimpanzees. The slight squabble was over and they were all feeding peacefully on some yellow plum-like fruits.

The distance between us was too great for me to make detailed observations, and I decided to try to get closer. Carefully I surveyed the trees close to the group: if I could manage to get to that large fig without frightening the chimpanzees, I thought, I would get an excellent view. It took me about ten minutes to make the journey. As I moved cautiously around the thick gnarled trunk of the fig I realised that the chimpanzees had gone; the branches of the fruit tree were empty. The same old feeling of depression clawed at me. Once again the chimpanzees had seen me and silently fled. Then all at once my heart missed several beats.

Less than twenty yards away from me two male chimpanzees were sitting on the ground staring at me intently. Scarcely breathing, I waited for the sudden panic-stricken flight which normally followed a surprise encounter between myself and the chimpanzees at close quarters. But nothing of the sort happened. The two large chimps simply continued to gaze at me. Very slowly I sat down and, after a few more moments, the two calmly began to groom one another.

As I watched, still scarcely believing it was true, I saw two more chimpanzee heads peering at me over the grass from the other side of a small forest glade; a female and a youngster. They bobbed down as I turned my head towards them, but soon reappeared, one after the other, in the lower branches of a tree about forty yards away. There they sat, almost motionless, watching me.

For over half a year I had been trying to overcome the chimpanzees' inherent fear of me, the fear which made them vanish into the undergrowth whenever I approached. At first, indeed, they had fled even when I was as far away as five hundred yards, and on the other side of a ravine. Now two males were sitting so close that I could almost hear them breathing.

Without the slightest doubt, this was the proudest moment I had known. I had been accepted by the two magnificent creatures grooming each other in front of me. I knew them both: David Greybeard, who had always been the least afraid of me, was one, and the other was Goliath, not the giant his name implies but of superb physique and the highest ranking of all the males. Their coats gleamed vivid black in the softening light of the evening.

For more than ten minutes David Greybeard and Goliath sat grooming each other and then, just before the sun vanished over the horizon behind me, David got up and stood staring at me. And it so happened that my elongated evening shadow fell across him. The moment is etched deep into my memory: the excitement of the first close contact with a wild chimpanzee and the freakish chance that cast my shadow over David even as he seemed to gaze into my eyes. Later it acquired an almost allegorical significance for, of all living creatures to-day, only man, with his superior brain, his superior intellect, overshadows the chimpanzee. Only man casts his shadow of doom over the freedom of the chimpanzee in the forests with his guns and his spreading settlements and cultivation. At that moment, however, I did not think of this. I only marvelled in David and Goliath themselves.

The depression and despair which had so often visited me during the preceding months were as nothing compared to the exultation that I felt when the group had finally moved away and I was hastening down the darkening mountain-side to my tent on the shores of Lake Tanganyika.

It had all begun three years before when I had met Dr Louis Leakey, the well-known anthropologist and palaeontologist, in Nairobi. Or perhaps it had begun in my earliest childhood. When I was just over one year old my mother gave me a toy chimpanzee, a large hairy model celebrating the birth of the first chimpanzee infant ever born in the London zoo. Most of my mother's friends were horrified and predicted that the ghastly creature would give a small child nightmares; but Jubilee (as the celebrated infant itself was named) was my most loved possession and accompanied me on all my childhood travels. Indeed, I still have the veteran toy to-day.

Quite apart from Jubilee, I had been fascinated by live animals from the time when I first learned to crawl. One of my earliest recollections is of the day that I hid in a small stuffy henhouse in order to see how a hen laid an egg. When I emerged, after about five hours, triumphant, the whole household had apparently been searching for me for hours, and my mother had even rung the police to report me missing.

It was about four years later, when I was eight, that I first decided that, when I grew up, I would go to Africa and live with wild animals. Although when I left school at eighteen I took a secretarial course and then two different jobs, the longing for Africa was still very much with me. So much so, indeed, that when I received an invitation to go and stay with a school friend at her parents' farm in Kenya, I handed in my resignation the same day and left a fascinating job at a documentary film studio in order to earn my fare to Africa by waitressing during the summer season in Bournemouth, my home town. It was cheaper to live at home.

'If you are interested in animals,' someone said to me about a month after my arrival in Africa, 'then you should meet Dr Leakey.' By then I had started on a somewhat dreary office job, for I had not wanted to overstay my welcome at my friend's farm. I was desperate to get closer to the animals which, for me, were Africa. I went to see Louis Leakey at the natural history museum where, at that time, he was curator. Somehow he must have sensed that my interest in animals was not just a passing phase, but was rooted deep, for on the spot he took me on as an assistant/secretary.

I learnt much whilst I worked at the museum: the staff were all keen naturalists full of enthusiasm and happy to share some of their boundless knowledge with me. Best of all, I was offered the chance, with one other girl, of accompanying Dr Leakey and his wife, Mary, on one of their annual palaeontological expeditions to Olduvai Gorge on the Serengeti plains. In those days, before the opening up of the Serengeti to tourists, before the discoveries of *Zinjanthropus* (the Nutcracker man) and *Homo habilis* at Olduvai, the area was completely secluded: the roads and tourist buses and light aircraft that pass there to-day were then undreamt of. It was a real expedition into the 'wilds of Africa,' such as I had dreamed of since childhood.

The digging itself was fascinating. For hours as I picked away at the ancient clay or rock of the Olduvai fault to extract the remains of creatures which had lived millions of years ago, the task would be purely routine. But from time to time, and without warning, I would be filled with awe by the sight or the feel of some bone I held in my hand. This – this very bone – had once been part of a living, breathing animal which had walked and slept and propagated its species. What had it really looked like? What colour was its hair; what was the odour of its body? Such questions filled my mind – questions which science will probably never be able to answer.

It was the evenings, however, that gave those few months their special enchantment for me. When the hard work of the day was finished, at about six o'clock, then Gillian, my fellow-assistant, and I were free to return to camp across the sun-parched arid plains above the gorge where we had sweated all day. Olduvai, in the dry season, becomes almost a desert, but as we walked past the low thorn bushes we often glimpsed dik-diks, those graceful miniature antelopes scarcely larger than a hare. Sometimes there would be a small herd of gazelles or giraffes and, on a few memorable occasions, we saw a black rhinoceros plodding along the gorge below. Once we came face to face with a young male lion: he was no more than forty feet away when we heard his soft growl and peered round to see him on the other side of a small bush. For the first time in my life I realised the full meaning of the phrase: 'my heart stood still.' We were down in the bottom of the gorge where the vegetation is comparatively thick in parts: slowly we backed away whilst he watched, his tail twitching. Then, out of curiosity, I suppose, he followed us as we walked deliberately across the gorge towards the open, treeless plains on the other side. As we began to climb upwards he vanished into the vegetation and we did not see him again.

It was during this period, I think, that Louis Leakey decided that I was the person for whom he had been searching for nearly twenty years – someone completely fascinated by animals and their behaviour; someone who could forego the amenities of civilisation for long periods of time without any difficulty. For it was towards the end of our time at Olduvai that he began to talk to me about a group of chimpanzees living on the shores of Lake Tanganyika.

The chimpanzee is found only in Africa where it ranges across the equatorial forest belt from the west coast to a point just east of Lake Tanganyika. The group Louis was referring to comprised chimpanzees of the Eastern or Long-haired variety, *Pan troglodytes schweinfurthi* as they are labelled by toxonomists. Louis described their habitat as mountainous, rugged and completely cut off from civilisation. He spoke, for a while, of the dedication and patience that would be required of any person who tried to study them.

Only one man, Louis told me, had attempted to make a serious study of chimpanzee behaviour in the wild; and Dr Henry Nissen, who had done this pioneering work, had only been able to spend two and a half months in the field – in French Guinea[1]. Louis said that no one could expect to accomplish much in such a short time – two years would scarcely be long enough. Much more Louis told me during that first talk. He was, he said, particularly interested in the behaviour of a group of chimpanzees living on the shores of a lake – for the remains of prehistoric man were often found on a lake shore and it was possible that an understanding of chimpanzee behaviour to-day might shed light on the behaviour of our stone-age ancestors.

As he talked I suppose I guessed what was coming, yet I could not believe that he spoke seriously when, after a pause, he asked me whether I would be willing to tackle the job. For, although it was the sort of thing I most wanted to do, I was, of course, completely unqualified to undertake a scientific study of animal behaviour.

Louis, however, knew exactly what he was doing. Not only did

1. Nissen. H. W. (1931). A field study of the Chimpanzee. *Comp. Psychol. Monogr.*, 8, 1-22.

he feel that a university training was unnecessary, but even that in some ways it might have been disadvantageous. He wanted someone with a mind uncluttered and unbiased by theory who would make the study for no other reason than a real desire for knowledge; and, in addition, someone with a sympathetic understanding of animals.

Once I had agreed wholeheartedly and enthusiastically to undertake the work, Louis embarked on the difficult task of raising the necessary funds. He not only had to convince someone of the need for the study itself, but also that a young and unqualified girl was the right person to attempt it. Eventually the Wilkie Foundation in Illinois, America, agreed to contribute a sum sufficient to cover the necessary capital expenses – a small boat, a tent and airfares – and an initial six months in the field. I shall always be immensely grateful to Mr Leighton Wilkie who, trusting in Louis's judgement, gave me the chance to prove myself.

By this time I was back in England, but as soon as I heard the news I made arrangements to return to Africa. The government officials in Kigoma in whose area I would be working, had agreed to my proposed study, but they were adamant on one score: they would not hear of a young English girl living in the bush alone without a European companion. And so my mother, Vanne Goodall, who had already been out to Africa for a few months, volunteered to accompany me on my new venture.

When we arrived in Nairobi everything at first went well. The Gombe Stream Chimpanzee Reserve (now the Gombe National Park), the home of my chimpanzee group, fell under the jurisdiction of the Tanganyika Game Department, and the Chief Game Warden was most helpful in sending the necessary permits for me to work in the Reserve. He also sent much useful information about the conditions there – the altitude and temperature, the type of terrain and vegetation, the animals I might expect to encounter. Word had come through that the little aluminium boat which Louis had bought had arrived safely in Kigoma. And Dr Bernard Verdcourt, Director of the East African Herbarium, volunteered to drive Vanne and I to Kigoma: he would be able to collect plant specimens on the way, and also in the botanically little-known Kigoma area.

Just as we were ready to leave came the first set-back. The District

Commissioner of the Kigoma region sent word that there was trouble amongst the African fishermen on the beaches of the chimpanzee reserve. The Game Ranger for the area had gone there to try and sort things out but, in the meantime, it would not be possible for me to begin my work.

Luckily for my peace of mind Louis immediately put forward the suggestion that I should make a short trial study of the vervet monkeys on an island in Lake Victoria. Within a week Vanne and I were on his motor launch chugging lazily over the shallow muddy water of the lake to uninhabited Lolue Island. With us were Hassan, captain of the little launch, and his assistant, both Africans of the Kakamega tribe. Hassan, who later joined me at the chimpanzee reserve, is a wonderful person. Always calm and a little stately, he is superb in an emergency and, with his sense of humour and intelligence, makes a fine companion. At that time he had worked for Louis for nearly thirty years.

It was three weeks before we received a radio message recalling us to Nairobi, and those three weeks were full of enchantment. At night we slept on the boat anchored just off the island and lulled by the gentle rocking swell of the lake. Every morning just before sunrise Hassan rowed me ashore in the little dinghy and I remained on the island watching the monkeys until dusk – or even later on those evenings when the moon was bright. Then I met Hassan on the lake shore and he rowed me back to the boat. Over our meagre supper, usually consisting of baked beans, eggs or tinned sausages, Vanne and I exchanged our news of the day.

This is not the place to tell of the sixteen monkeys of my little troop and their fascinating behaviour, for I am writing a book about chimpanzees. But the short study taught me a good deal about such things as note-taking in the field, the sort of clothes to wear, the movements which a wild monkey will tolerate in a human observer and those it will not. And, whilst the chimpanzees reacted quite differently in many ways, nevertheless, the things I learnt at Lolue were very helpful when I started work at the Gombe Stream.

I was sorry, in a way, when the expected message came one evening, for it meant leaving the vervets just as I was beginning to learn about their behaviour, just as I had become familiar with the different individuals of the troop. It is never nice to leave a job unfinished. Once we reached Nairobi, however, I could think of nothing save the excitement of the eight hundred-mile journey to Kigoma – and the chimpanzees. Nearly everything had been ready before we left for Lolue, so it was only a few days before we were able to set off with Bernard Verdcourt for Kigoma.

The journey itself was fairly uneventful, although we had three minor breakdowns, and the Land-Rover was so badly overloaded with all our equipment that it swayed dangerously if we went too fast. When we reached Kigoma, however, after a dusty three days on the road, we found the whole town in a state of chaos: since we had left Nairobi violence and bloodshed had erupted in the Congo which lay only some twenty-five miles to the west of Kigoma, on the other side of Lake Tanganyika. Kigoma was overrun by boat-loads of Belgian refugees. It was Sunday when we drove, for the first time, down the avenue of mango trees that shades Kigoma's one main street. Everything was closed, and we could find no official to help us.

Eventually, however, we ran the District Commissioner to earth, and he explained, regretfully but firmly, that there was no chance at all of my proceeding to the chimpanzee reserve. First it was necessary to wait and find out how the local Kigoma district Africans would react to the tales of rioting and disorder in the Congo. It was a bitter blow, but there was little time for moping.

We booked ourselves in a room each at one of the two hotels, but this luxury did not last long. Another boat-load of refugees arrived that evening, and every available inch of space was needed. Vanne and I doubled up, squeezing ourselves into the small amount of room that was left over after we had packed in all our equipment from the Land-Rover. Bernard shared his room with a couple of homeless Belgians, and we even got out our three camp beds and lent them to the harassed hotel owner. Every room was crammed, but these refugees were in paradise compared to those temporarily housed in the huge warehouse, normally used for storing cargo on its way across the lake to or from the Congo. There everyone slept in long rows on mattresses or merely blankets on the cement floor, and queued up in their hundreds for the meagre meals that Kigoma was able to provide for them.

Very soon Vanne, Bernard and I had made the acquaintance of a number of Kigoma's residents. Of course, we offered to help with the catering and this offer was eagerly accepted. On our second evening in Kigoma we three and a few others made two thousand spam sandwiches. These were finally stacked neatly in wet cloths in large tin trunks which were carted off to the warehouse. Later we helped to hand them out to the refugees, together with soup, some fruit, chocolate, cigarettes and drinks. I have never been able to face tinned spam from that day to this.

Two evenings later most of the refugees had gone, carried off by a series of extra trains to Tanganyika's capital, Dar es Salaam. The hustle of activity was over, but still we were not allowed to leave for the chimpanzee reserve. We all became somewhat depressed. My funds did not permit Vanne and I to stay on at the hotel, so we decided to put up a temporary camp somewhere. When we inquired where we could do this, we were directed to the grounds of the Kigoma prison! Actually this was not as bad as it sounds, for the grounds, which are beautifully kept, overlook the lake and, at that time of year, the citrus trees all around were groaning under the weight of sweet-smelling oranges and tangerines. The mosquitoes in the evening were terrible though.

During our period of enforced inactivity we got to know the tiny town of Kigoma quite well – it is more like a village by European or American standards. The hub of activity was down by the lake shore where the natural harbour offers anchorage to the boats which ply up and down the lake to Burundi, Zambia, Malawi and across to the Congo in the west. Near the lake, too, are the government administrative offices, the police station, the railway station and the post office.

One of the most fascinating aspects of any small town in Africa is the colourful fruit and vegetable market where the merchandise is offered for sale in small piles, each of which has been accurately counted and priced. In Kigoma market we found that the more prosperous traders operated from under a lofty stone awning; the others sat on the red earth of the main market square, their wares neatly set out on sacking or on the ground itself. Bananas, green and yellow oranges, and dark purple, wrinkled passion fruits were displayed in profusion, and there were bottles and jars of glowing red cooking oil made from the fruit of oil nut palms.

Kigoma boasts one main street which slopes upward from the administrative centre and runs through the main part of Kigoma. On either side it is flanked by tall shady mango trees and countless

23

tiny stores or *dukas* as they are called throughout East Africa. It amazed us, as we walked through Kigoma, that so many stores could survive when they all appeared to sell similar goods. Again and again we saw piles of kettles and crockery, plimsolls and shirts, torches and alarm clocks. Most stores were brightened by great squares of brilliantly coloured material sold in pairs to the African women and known as Kangas. One square is wrapped around under the arms and hangs down just below the knees; the other becomes a headdress. Outside some of the *dukas* a tailor worked at his foot-operated sewing machine, and an old Indian sat in the dust outside the tiny shoe shop using his feet like extra hands to hold the leather as he sewed and tacked and glued shoes together. He was so skilful that it was a delight to watch him.

We got better acquainted with several Kigoma residents during these days; they were mostly government officials and their wives, and very charming and hospitable we found them. I shall never forget when Vanne, not wanting to rebuff any of our new-found friends, accepted two offers of a hot bath for the same evening. Bernard, who was convinced that we were both slightly mad anyway, drove her stoically from house to house to keep her appointments without giving her away.

When we had been in Kigoma just over a week, David Anstey, the Game Ranger who had been sorting out the troubles between the fishermen at the Gombe Stream Reserve, returned to Kigoma. He and the District Commissioner had a long conference, the outcome of which was that I was given official permission to proceed to the Gombe Stream. By this time I had almost given up hope of even seeing a chimpanzee for I had convinced myself that, at any minute, we would be ordered back to Nairobi. And so, when I found myself on the government launch which had been lent to us for transporting all our equipment, including our twelve-foot dinghy, the expedition had taken on a dreamlike quality.

As the engine sprang to life and the anchor was drawn, we waved good-bye to Bernard and were soon steaming out of Kigoma harbour and turning northward along the eastern shores of the lake. I can remember looking down into the incredibly clear water and thinking to myself: 'I expect the boat will sink, or I shall fall overboard and be eaten by a crocodile.' But neither of these things happened.

I would follow the chimpanzees through the forest

Inquisitive African children visit our camp (*Copyright National Geographic Society*)

With her clinic Vanne established friendly relations with the fishermen

*Chapter 2*  Early Days

I retained the strange feeling that I was living in a dream world throughout the twelve-mile journey from Kigoma to our camping place in the Gombe Stream Reserve. It was the middle of the dry season, and the shore-line of the Congo, though it was only twenty-five miles distant, was not even faintly discernible to the west of long narrow Lake Tanganyika. The fresh breeze and the deep blue of the water, which was choppy with little waves and flecked with white foam, combined to make us feel that we were at sea.

I gazed at the eastern shore-line. Between Kigoma and the start of the chimpanzee sanctuary the steep slopes of the rift escarpment, which rise 2500 feet above the lake, are, in many places, bare and eroded from years of tree felling. In between, small pockets of forest cling to the narrow valleys where fast-flowing mountain streams rush down to the lake. The coast-line is broken into a series of elongated bays often separated by rocky headlands that jut out into the lake. We steered a straight course, proceeding from headland to headland, but we noticed that the little canoes of the fishermen hugged the shore-line. David Anstey, who was travelling with us to introduce us to the local African inhabitants of the area, explained that sometimes the lake becomes suddenly rough, with fierce winds sweeping down the valleys and churning the water into a welter of spray and waves.

All along the shore-line tiny fishing villages clung to the mountain slopes or nestled in the mouths of the valleys. The dwellings were mostly simple mud and grass huts, although, even in those days, there were a few larger buildings roofed with shiny corrugated iron – that curse, for those who love natural beauty, of the modern African landscape.

When we had travelled about seven miles, David pointed out the large rocky outcrop that marks the southern limits of the chimpanzee reserve. Once past the boundary we noticed that the country had changed suddenly and dramatically: the mountains were thickly wooded and intersected by valleys supporting dense tropical forests. Even here we noticed a number of fishermen's huts dotted at intervals along the white beaches: David explained that these were temporary structures. The Africans had permits to fish during the dry season, and to dry their catch in the sun on the beaches of the Reserve itself. But when the rains began the fishermen returned to their home villages outside the chimpanzee reserve. It was amongst these men that the recent trouble had broken out – some argument between the fishermen of two different villages as to which of them had the right to a certain stretch of beach.

Since that day I have often wondered exactly what it was that I felt as I stared at the wild country that, so soon, I should be roaming. Vanne admitted afterwards that she was secretly horrified by the steepness of the slopes and the impenetrable appearance of the valley forests, and David Anstey told me, several months later, that he had guessed I would be packed up and gone within six weeks. I remember feeling neither excitement nor trepidation, but only a curious sense of detachment. What had I, the girl standing on the government launch in her jeans, to do with the girl who, in a few days, would be searching those very mountains for wild chimpanzees? Yet by the time I went to sleep that night, the transformation had taken place.

After a two-hour journey the launch dropped anchor at Kasakela where the two government Game Scouts had their headquarters. David Anstey had suggested that, at least until we were familiar with the area, we should camp somewhere near their huts. As our dinghy approached the white sandy beach we saw that quite a crowd had gathered to watch our arrival: the two scouts, the few Africans who had permission to live permanently at the Reserve so that the scouts should not be completely isolated, and some of the fishermen from nearby huts. We stepped ashore, splashing into the sparkling wavelets, and were greeted first by the scouts and then, with great ceremony, by the honorary headman of Kasakela village, old Iddi Matata. He was a colourful figure with his red turban, red European-style coat over flowing white robes, wooden shoes and white beard. He made a long speech of welcome to us in Swahili, of which I understood only fragments, and we presented him with a small gift which David had advised us to buy for him.

The formalities over, Vanne and I followed David for about thirty yards along a narrow track leading from the beach through thick vegetation to a small natural clearing. With the help of David and the African scouts, the large tent which Vanne and I would share was soon up. Behind the tent flowed a small gurgling stream, and the clearing was shaded by tall oil nut palms. It was a perfect camp site. Fifty yards or so distant, under some trees on the beach, we set up another small tent for Dominic, the cook we had employed before leaving Kigoma.

When our camp was organised I slipped away to explore. The tall grass of the lower mountain slopes had been burnt off by a recent bush fire, and the charcoal-encrusted ground was slippery. It was about four o'clock, but still the sun burned down fiercely and I was hot and sweating by the time I had climbed high enough to look down over the flat lake and the broad valley, lush and green by contrast with the blackened mountain-side on which I stood.

I sat on a big flat rock, hot from the sun, and slowly I could feel myself coming alive after the depression of that week in Kigoma and the trance-like state in which I had been since leaving that morning. Presently a troop of about sixty baboons passed by, searching the newly burnt ground for the remains of baked insects. Some of them climbed trees when they saw me and shook the branches with jerky threatening movements, and two of the big males started their loud alarm bark. But on the whole the troop was not very worried by my presence and soon moved slowly on, busy with its own affairs. I saw a bushbuck too, a graceful chestnut animal, slightly larger than a goat, with sturdy spiralling horns. He stared at me, motionless, and then suddenly turned and bounded away, barking like a dog and flashing the white underside of his tail.

I only stayed out on the mountain about three-quarters of an hour, but when I returned, almost as black as the slopes on which I had been scrambling, I no longer felt an intruder. That night I pulled my camp-bed into the open and slept with the stars above me, twinkling down through the rustling fronds of a palm tree.

The next morning, of course, I was eager to go out looking for chimpanzees; but I soon found that, to start with anyway, I was not to be my own master. David Anstey had arranged for a number

27

of the local Africans to come and meet Vanne and me. He explained that they were all worried and resentful; they could not believe that a young girl would come all the way from England just to look at apes, and so the rumour had spread that I was a government spy. I was, of course, very grateful to David for sorting things out for me, right at the start, but my heart sank lower and lower when I heard the plans he had made for me.

Firstly it was agreed that the son of the chief of Mwamgongo, a large fishing village to the north of the chimpanzee reserve, should accompany me. He would make sure that, when I saw one chimpanzee, I did not write down in my book that I had seen ten or twenty. I realised later that the Africans were still hoping to reclaim the thirty square miles of the Reserve for themselves: if I stated that there were more chimpanzees than, in fact, there were, then, the Africans felt, government could make a better case for keeping the area a protected reserve. Secondly, David felt that, for the sake of my prestige, I should employ an African to carry my haversack.

Since I was convinced that the only way to establish contact with shy animals was to move about alone, I was upset to think that I must be encumbered by two companions. Then came the final blow – I also had to have one of the Game Scouts with me. I felt depressed and miserable when I went to bed that night.

When I woke up the next morning, however, everything was new and exciting and my gloom soon vanished. I had arranged to meet the chief's son at a valley near the northern boundary, for my Game Scout, Adolf, had reported seeing chimpanzees whilst patrolling there the day before. David Anstey had some business in Mwamgongo village so he took me, together with Adolf and Rashidi, my 'porter,' in his boat and dropped us off at the appointed place.

The chief's son came towards us, followed by five or six other Africans, and I had a sudden horror that they would all insist on coming with me. But my fears were quickly allayed. The young man asked where I planned to go and I pointed vaguely towards the steep thickly forested slopes of the valley. He looked rather taken aback and began to talk quietly and earnestly with his friends in Kiha, the language of the local Ha tribesmen. A few moments later he again approached and told me that he felt rather unwell and would not go with me that day. Later I found out that he had expected that I would merely ride up and down the lake-shore in a boat counting any chimpanzees I saw. The idea of clambering about in the mountains did not appeal to him at all, and I never saw him again.

Just as we were leaving, two fishermen ran up and asked us to follow them for a moment. They led us to a tree just behind one of the temporary huts: the bark was gashed in a hundred places. There, we learned, a lone bull buffalo had chased one of the fishermen the night before. The man had managed to climb to safety but the buffalo had charged at the slender tree time and time again. Whether the men were merely making a report to the Game Scout, or whether they were trying to impress upon me the hazards of their country, I do not know, but the memory of that battered tree haunted me for weeks when I was crawling about in the thick undergrowth of the valley forests.

After this we set off up the Mitumba Valley, and I soon found myself walking through the sort of African forest of which I had always dreamed. There were giant buttressed trees festooned with lianas and, here and there, brilliant red or white flowers that gleamed through the dark foliage. We moved along beside a fast-flowing shallow stream, often wading across when the going was easier on the opposite side. Now and then a kingfisher or some other forest bird flashed past, and once a troop of red-tail monkeys leapt across a gap overhead, their coppery tails glinting. The forest canopy, over a hundred feet above us, shut out most of the sunlight and there was little undergrowth along the valley bottom.

When we had been travelling for about twenty minutes Adolf left the stream and led us up the side of the valley. Almost at once the going became much more difficult: the trees were smaller and the undergrowth was thick and tangled with vines so that we had to crawl most of the time. Presently Adolf stopped under an enormous tree. I looked up and saw that it was laden with small orange and red fruits: down below the ground was littered with broken-off twigs and partly eaten fruits. This was the msulula tree where the chimpanzees had been feeding the day before.

I had not intended to go right up to the tree and, hoping that we had not disturbed any chimpanzees, I quickly told my guides that I wanted to watch from farther away – from the other side of the valley.

Ten minutes later we were settled in a little grassy clearing at about the same height as the msulula tree and directly opposite. I discovered, afterwards, that it was, in fact, the only really good open space from which one could watch the tree: Rashidi's trained eye had spotted it instantly whereas I, at that time, would never have noticed it. It seemed very quiet and peaceful away from the rushing of the stream. Some cicadas were shrilling incessantly, and there were some bird songs and the occasional bark or scream of a baboon.

All at once I was tense with excitement for, in the valley below, I heard the calling of a group of chimpanzees. I had heard chimpanzees in the zoo, of course, but out here, in the African forest, the sound was thrilling beyond words. First one chimp gave a series of low, resonant 'pant-hoots' – loud hooting calls connected by audible inhalations of breath. These grew louder and louder until, in the end, the chimpanzee was almost screaming. Half-way through his calling another one joined in and then another. I had read of chimpanzees drumming on tree-trunks when I went through the report of Dr Nissen: now I heard the strange, echoing sound for myself, reverberating through the valley, interspersed with the wild chorus of pant-hoots.

The group was very close to the msulula tree and I waited, with every sense alert, staring into the forest opposite. Even so it was Rashidi who saw the first movement as a chimpanzee climbed up a palm trunk and so into the branches of the giant tree. It was followed by another and another and another, each climbing in orderly procession. I counted sixteen in all, some large, some much smaller. One was a mother with a tiny infant clinging to her tummy.

Excited as I was I could not help feeling disappointed for, although the chimpanzees remained for two hours in the tree, I saw little except an occasional glimpse of a black arm reaching out from the foliage and pulling bunches of fruit out of sight. And then, one after the other and in complete silence, the whole group climbed down the palm tree ladder and vanished into the forest. That was what amazed me most – sixteen chimpanzees in one tree and yet the only sounds I had heard them make had been the calls announcing their arrival.

A few minutes after the last chimpanzee had gone, when I was still searching the valley with my binoculars in the hope of seeing the group climbing into another tree, I was horrified to hear Adolf and Rashidi announce that we must return as it was their time for eating. I remonstrated in vain: until I felt myself on firmer ground I dared not order them to stay, nor did I dare risk David Anstey's displeasure by remaining without my escorts. As we filed back through the forest, however, I determined that things would be different in future.

The msulula continued to bear fruit for another ten days and Rashidi and Adolf took turns to accompany me, taking their midday meals with them. For three nights, in fact, we slept there, Rashidi and Adolf huddled close together in the glow of a small camp-fire, and myself, farther back, wrapped up in a blanket.

I saw many chimpanzees during those ten days. Sometimes large groups climbed up and fed on the msulula fruits, and sometimes there were only two or three individuals in a group. Twice I saw a completely solitary male who fed for over an hour by himself. I soon realised that the groups were not stable: once, for instance, fourteen chimps arrived together but they left in two different groups, the second climbing down a full half-hour after the first. And, judging from the subsequent sounds, they went off in different directions. On another occasion I saw two small groups meet up in the tree, with much screaming and rushing about through the branches. Then they quietened, fed peacefully together and, so far as I could tell, moved off together. I found out too that some groups comprised only adult males, some only females and youngsters, and others consisted of males, females and youngsters all together.

I was, however, far from satisfied. I saw very few interactions between individual chimpanzees – the foliage of the msulula was far too thick – and when I twice tried to watch from nearer by I failed dismally: once the chimps saw me on their way to the tree and fled, and on the second occasion all I saw of the four who fed for an hour almost above me was a very brief glimpse as they climbed up, and subsequently down, the palm tree ladder.

Later, however, I realised how lucky I had been during the fruiting of the msulula tree: I probably learnt more during those ten days than I did during the eight depressing weeks that followed. Search as we would, we found no other large fruiting tree or group of fruiting trees. We went up most of the twelve valleys of the

Reserve, but the undergrowth was often very thick and, whilst the sound of the streams certainly drowned any noise we made, it also effectively camouflaged the sounds which might have warned us of the whereabouts of chimps. Those which we did see were usually so close by the time we came upon them that they fled instantly. I can well imagine, now, how many times they must have seen us coming and silently vanished without our ever being aware of their presence.

We had slightly better luck when we climbed the ridges between the valleys but, again, the chimps either hurried away at the sight of us, even when we were five hundred yards distant on the other side of a ravine, or else were so far off that it was impossible to see any details of their behaviour. For a while I wondered whether it was because there were three of us that the chimpanzees seemed so frightened, but even when I left my two companions on some high peak, from where they could keep an eye on me, and tried to get closer to some distant group on my own, the reaction of the chimps was the same. They fled.

In between the disappointing days when we only saw chimps too far off to observe properly or, for a few minutes, close by, before they fled, were even worse days when we saw no chimps at all. The more I thought of the task I had set myself, the more despondent I became. Nevertheless, those weeks did serve to acquaint me with the rugged terrain. My skin became hardened to the rough grasses of the valleys, and my blood immune to the poison of the tsetse fly so that I no longer swelled hugely each time I was bitten. I became increasingly sure-footed on the treacherous slopes that were equally slippery whether they were bare and eroded, crusted with charcoal, or carpeted by dry, trampled grass. Gradually, too, I became familiar with many of the animal tracks in the five valleys that became my main work area.

I encountered many of the other denizens of the mountains during our daily wanderings: huge grey bushpigs with their silvery spinal crests; troops of banded mongooses rustling through the leaves in search of insects; the squirrels and the striped and spotted elephant shrews of the thick forests. Gradually, too, I sorted out the many different kinds of monkeys which can be found in the Gombe Stream area. Most often we came across troops of baboons: sometimes they accepted our presence quite calmly, like

Chimpanzees do not shelter from the rain

those I met that first afternoon, but others would bark persistently and noisily until we, or they, had moved out of sight. In each of the valleys there were two or three small troops of red-tail monkeys and a few blue monkeys. There were much larger groups of red colobus monkeys, numbering sixty or more, each of which ranged over a couple of valleys. Occasionally we came across a solitary silver monkey with its black face framed by a white band of hair. There were even a few troops of vervet monkeys, down by the lake-shore, to remind me of my monkey-watching days on Lolue Island.

In particular I liked to watch the red colobus. They are large monkeys, and occasionally I fleetingly mistook the adult males for chimpanzees because, in some lights, the dark brown hair of their backs looks black, and they sit upright on the branches, holding an overhead bough with one hand, more like apes than monkeys. But their long, thick hanging tails soon revealed their true identity. When I got close to them and they peered down at me from the branches, their faces always reminded me of surprised maiden aunts wearing ginger-red golliwog wigs.

Rashidi taught me a great deal about bush lore and how to find my way through seemingly impenetrable forest and, despite my initial disappointment when I had heard that I was not allowed to be alone, I was grateful for his help during those early days. But he soon had to leave me, to return for a while to his village, and, as I found that Adolf the scout was not suited to long arduous hours without food in the mountains, I had a succession of other African companions during the next few months. There was Soko, from Nyanza, whose name caused much amusement amongst the local Africans, for this is their name for the chimpanzee! Next came the enormously tall and willowy Wilbert who always looked immaculate even after scrambling on his belly along a pig trail; and finally Short who, as his name implies, was very small. All three were tough men who had spent their lives working in the bush with animals, and I enjoyed their company and learnt much during the days they worked for me.

The rainy season lasts for six months

## Chapter 3

First Observations

About three months after our arrival, Vanne and I fell ill at the same time. It was undoubtedly some sort of malaria but, as we had been told by no lesser person than the doctor in Kigoma that there was no malaria in the area, we had no drugs with us. How he came to believe such a strange fallacy I cannot imagine, but we were too naïve to question him at the time. For nearly two weeks we lay, side by side, on our low camp-beds in our hot stuffy tent sweating out the fever. Occasionally we mustered the strength to take our temperatures – there was nothing else to do to pass the time for neither of us felt like reading. Vanne had a temperature of 105° almost constantly for five days – it dropped slightly only during the coolness of the nights. Afterwards we were told that she had been lucky to pull through at all. To make everything worse, the whole camp was pervaded, throughout our illness, by the most terrible smell – rather like bad cabbage water. It was the flower of some tree – I forget its name now – I always think of it as the 'fever flower tree.'

Dominic, our cook, was wonderful during those days. He begged us to go into Kigoma to see a doctor, and when we pleaded that we felt much too ill to face the three-hour journey in our little boat, he made up for our lack of medical attention by constantly fussing over us. One night Vanne wandered out of the tent in a delirium and fell, unconscious, by one of the palm trees. I never knew she had left the tent. It was Dominic who found her, at about three in the morning, and assisted her back to bed. Later he told us that he came along several times each night to make sure his 'Memsahibs' were all right.

As soon as the fever left me I was impatient to start work again. Nearly three months had sped away and I felt that I had learned nothing: I was frantic, for in a couple of months my funds would run out. I could not bear the thought of any of my African companions seeing me in my weak state and so, risking official displeasure, I set off alone one morning to climb the mountain I had climbed on my first afternoon – the mountain which rose directly above our camp. I left at my usual time, when it was still cool, in the first faint glimmerings of dawn. After ten minutes or so my heart began to hammer wildly, I could feel the blood pounding in my head, and I had to stop to catch my breath. Eventually, however, I reached an open peak about one thousand feet above the lake and, as it offered a superb view over the home valley, I decided to sit there for a while and search for signs of chimpanzees through my binoculars.

I had been there some fifteen minutes when a slight movement on the bare burnt slope just beyond a narrow ravine caught my eye. I looked round and saw three chimps standing there staring at me. I expected them to flee, for they were no farther than eighty yards away, but after a moment they moved on again, quite calmly, and were soon lost to sight in some thicker vegetation. Had I been correct, after all, in my assumption that they would be less afraid of one person, completely alone? For, even when I had left my African companions behind and approached a group on my own, the chimps had undoubtedly been fully aware of what was going on.

I remained on my peak and later on in the morning a group of chimps, with much screaming and barking and pant-hooting, careered down the opposite mountain slope and began feeding in some fig trees that grew thickly along the stream banks in the valley below me. They had only been there about twenty minutes when another procession of chimps crossed the bare slope where, earlier, I had seen the three. This group also saw me – for I was very conspicuous on the rocky peak. But, although they all stopped and stared and then hastened their steps slightly as they moved on again, the chimpanzees did not run in panic. Presently, with violent swaying of branches and wild calling, this group joined the chimpanzees already feeding on figs. After a while they all settled down to feed quietly together, and when they finally climbed down from the trees they moved off in one big group. For part of the way, as they walked up the valley, I could see them following each other in a long, orderly line. Two small infants were perched, like jockeys, on their mothers' backs. I even saw them pause to drink, each one for about a minute, before leaping across the stream.

It was by far the best day I had had since my arrival at Gombe, and when I got back to camp that evening I was exhilarated, if exhausted. Vanne, who had been far more ill than I and who was still in bed, was much cheered by my excitement.

That day, in fact, marked the turning-point in my study. The fig trees grow all along the lower reaches of the stream and, that year, the crop in our valley was plentiful and lasted for eight weeks. Every day I returned to my peak, and every day chimpanzees fed on the figs below. They came in large groups and small groups, singly and in pairs. Regularly they passed me, either moving along the original route across the open slope just above me, or along one or other of the trails crossing the grassy ridge below me. And, because I always looked the same, wearing similar dull-coloured clothes, and because I never tried to follow them or harass them in any way, the shy chimpanzees began to realise, at long last, that I was not, after all, so horrific and terrifying. Of course, for the most part I was completely alone; there was no need for my African companions to follow me up and down since they knew where I was going to be. When Short had to leave I decided to employ no other African and, although Adolf and, afterwards, Saulo David, the new scout, often came up in the evenings to make sure I was all right, for the most part I was completely on my own.

My peak quickly became *the* Peak. It is, I think, the very best vantage point for watching chimpanzees in the whole of the Gombe Stream sanctuary. Of course, from higher up there is a magnificent view in all directions, but the chimpanzees seldom move about near the top of the rift escarpment for most of their food is lower in the mountains. From the Peak I was able to look southward over our home valley, and also, if I walked just a few yards to the north, I could look down into the basin of lower Kasakela Valley, a thick almost circular pocket of forest. I quickly found that it was easy to cross the upper Kasakela Valley almost on the level through a fairly open wood – where, on several occasions, I came across a small herd of about sixteen buffalo. To the north of Buffalo Wood another open ridge offered a good view over the upper reaches of the narrow, steep-sided Mlinda Valley.

I carried a small tin trunk up to the Peak and there I kept a kettle, some coffee, a few tins of baked beans, a sweater and a blanket. A tiny stream trickled through Buffalo Wood. It was almost non-existent in the dry season, but I scooped out a shallow bowl in the gravelly stream bed and was able to collect enough of the sparkling clear water for my needs. So, when the chimpanzees slept near the Peak, I often stayed up there too – then I didn't have to trudge up the mountain in the morning. I was able to send messages down to Vanne with whichever of the Game Scouts climbed to the Peak in the evening so that she always knew when I was planning to stay out for the night.

For about a month I spent most of each day either on the Peak or overlooking Mlinda Valley where the chimps, before or after stuffing themselves with figs, ate large quantities of small purple fruits which tasted, like so many of their foods, as bitter and astringent as sloes or crab apples. Piece by piece, I began to form my first somewhat crude picture of chimpanzee life.

The impression that I had gained when I watched the chimps at the msulula tree of temporary, constantly changing associations of individuals within the community was substantiated. Most often I saw small groups of four to eight moving about together. Sometimes I saw one or two chimpanzees leave such a group and wander off on their own or join up with a different association. On other occasions I watched two or three small groups joining up to form a larger one.

Often, as one group crossed the grassy ridge separating the Kasakela Valley from the fig trees in the home valley, the male chimpanzees of the party would break into a run, sometimes moving in an upright position, sometimes dragging a fallen branch, sometimes stamping or slapping on the hard earth. These charging displays were always accompanied by loud pant-hoots, and afterwards the chimpanzee often swung up into a tree overlooking the valley he was about to enter, and sat quietly, peering down and obviously listening for a response from below. If there were chimps feeding in the fig trees they nearly always hooted back, as though in answer. Then the new arrivals hurried down the steep slope and, with more calling and screaming, the two groups met up in the fig trees. When groups of females and youngsters, with no males present, joined other feeding chimpanzees there was usually none of the excitement; the newcomers merely climbed up into the trees, greeted some of those already there, and began to stuff themselves with figs.

Whilst many details of their social behaviour were hidden from me by the foliage, I did get occasional fascinating glimpses. I saw one female, newly arrived in a group, hurry up to a big male and hold her hand towards him. Almost regally he reached out, clasped her hand in his, drew it towards him and kissed it with his lips. I saw two adult males embrace each other in greeting. I saw youngsters having wild games through the tree-tops, chasing around after each other or jumping again and again, one after the other, from a branch to a springy bough below. I watched small infants dangling happily by themselves for minutes on end, patting at their toes with one hand, rotating gently from side to side. Once two tiny infants pulled on opposite ends of a twig in a gentle tug-of-war. Often, during the heat of midday, or after a long spell of feeding, I saw two or more adults grooming each other, intently looking through the hair of their companions.

At that time of year the chimps usually went to bed late, making their nests when it was too dark to see properly through binoculars, but sometimes they nested earlier so that I could watch them from the Peak. I found that every individual, except for infants who slept with their mothers, made his own nest each night. Usually this took about three minutes: the chimp chose a firm foundation, such as an upright fork or crotch, or two horizontal branches. Then he reached out and bent over smaller branches on to this foundation, keeping each one in place with his feet. Finally he tucked in the little leafy twigs growing round the rim of his nest and then lay down. Quite often a chimp sat up after a few minutes and picked a handful of leafy twigs which he put under his head or some other part of his body before settling down again for the night. One young female I watched went on and on bending down branches until she had constructed a huge mound of greenery on which she finally curled up.

I climbed up into some of the nests after the chimpanzees had left them – most of them were built in trees that, for me, were almost impossible to climb. I found that there was quite compli-cated interweaving of the branches in some of them. I found, too, that the nests were never fouled with dung – and later, when I was able to get closer to the chimps, I saw how they were always careful to defecate and urinate over the edge of their nests, even in the middle of the night.

*     *     *

It was during that month that I really got to know the country well, for I often went on expeditions from the Peak, sometimes to examine nests, more often to collect specimens of the chimpanzees' food plants which Bernard Verdcourt had kindly offered to identify for me. Soon I could find my way around the sheer ravines and up and down the steep slopes of three valleys – the home valley, the Pocket, and Mlinda Valley – as well as a taxi-driver finds his way about in the highways and byways of London. It is a period I remember vividly, not only because I was beginning to accomplish something at last, but also because of the delight I felt in being completely by myself. For those who love to be alone with nature I need add nothing further; for those who do not, no words of mine could ever convey, even in part, the almost mystical awareness of beauty and eternity that accompany certain treasured moments. And, though the beauty was always there, those moments came upon me unaware: when I was watching the pale flush preceding dawn; or looking up through the rustling leaves of some giant forest tree, into the greens and browns and black shadows that occasionally ensnared a bright fleck of the blue sky; or when I stood, as darkness fell, with one hand on the still warm trunk of a tree and looked at the sparkling of an early moon on the never-still, sighing water of the lake.

One day, when I was sitting by the little trickle of water in Buffalo Wood, pausing for a moment in the coolness before returning from a scramble in Mlinda Valley, I saw a female bushbuck moving slowly along the nearly-dry stream bed. Occasionally she paused to pick off some plant and crunch it up. I kept absolutely still, and she was not aware of my presence until she was little more than ten yards away. Suddenly she tensed and stood, one small forefoot raised, staring at me. Because I did not move, she did not know what I was – only that my outline was somehow strange. I saw her velvet nostrils dilate as she sniffed the air, but I was down-wind and her nose gave her no answer. Infinitely slowly she came closer and closer, one step at a time, her neck craned forward, always poised for instant flight. I can still scarcely believe that her nose actually touched my knee: yet if I close my eyes I can feel again, in imagination, the warmth of her breath and the silken impact of her skin. But suddenly I blinked and she was gone in a flash, bounding away with loud barks of alarm until the vegetation hid her completely from my view.

It was rather different when, as I was sitting on the Peak, I saw a leopard coming towards me, his tail held up straight. He was at a slightly lower level than I, and obviously had no idea I was there. Ever since I arrived in Africa I had had an ingrained, illogical fear of leopards. Already, whilst working at the Gombe, I had several times nearly turned back when, as I crawled through some thick undergrowth, I had suddenly smelt the rank smell of cat. I had forced myself on, telling myself that my fear was foolish, that only wounded leopards charged humans with savage ferocity.

On this occasion, though, the leopard presently went out of sight as it started to climb up the hill – the hill on the peak of which I sat. I quickly hastened to climb a tree but, half-way there, I realised that leopards could climb trees. So I uttered a sort of half-hearted squawk. The leopard, my logical mind told me, would be just as frightened of me if he knew I was there. Sure enough, there was a thudding of suddenly startled feet and then silence. I returned to the Peak, but the feeling of unseen eyes watching me was too much. I decided to watch for chimps in Mlinda Valley for a while! And, when I returned to the Peak several hours later, there, on the very rock which had been my seat, was a neat little pile of leopard dung. He must have watched me go and then, very carefully, examined the place where such a frightening creature had been, and tried to exterminate my horrible scent with his own!

* * *

As the weeks went by the chimpanzees became less and less afraid. Quite often when I was on one of my food-collecting expeditions I came across chimpanzees unexpectedly, and after a while I found that some of them would tolerate my presence provided they were in fairly thick forest, and I sat quite still and did not try to move closer than sixty to eighty yards. And so, during my second month of watching from the Peak, when I saw a group settle down to feed I sometimes moved closer and was thus able to make more detailed observations.

It was at this time that I began to recognise a number of different individuals. As soon as I was sure of knowing a chimpanzee if I saw it again, I named it. Some scientists feel that animals should be labelled by numbers – that to name them is anthropomorphic – but I have always been interested in the *differences* between individuals and a name is not only more individual than a number, but also far easier to remember. Most names were simply those which, for some reason or other, seemed to suit the individuals to whom I attached them. A few chimps were named because some facial expression or mannerism reminded me of human acquaintances.

The easiest individual to recognise was old Mr McGregor. The crown of his head, his neck and his shoulders were almost entirely devoid of hair, but a little frill remained around his head rather like a monk's tonsure. He was an old male – perhaps between thirty and forty years of age (the longevity record for a captive chimp is forty-seven years). During the early months of my acquaintance with him, Mr McGregor was somewhat belligerent. If I accidentally came across him at close quarters he would threaten me, with an upward and backward jerk of his head and a shaking of branches, before climbing down and vanishing from my sight. He reminded me, for some reason, of Beatrix Potter's old gardener in *The Tale of Peter Rabbit*.

Ancient Flo, with her deformed, bulbous nose and ragged ears, was equally easy to recognise. Her youngest offspring at that time was two-year-old Fifi, who still rode everywhere on her mother's back, and Flo's juvenile son, Figan, was always to be seen wandering around with his mother and little sister. He was then about six years old; it was approximately a year before he would attain puberty. Flo often travelled with another old mother, Olly. Olly's long-shaped face was also distinctive: the fluff of hair on the back of her head – though no other feature – reminded me of my aunt, Olwen. Olly was also accompanied by two children, a daughter younger than Fifi, and an adolescent son, about a year older than Figan.

Then there was William who, I am certain, must have been Olly's blood-brother. I never saw any special signs of friendship between them, but their faces were amazingly alike. They both had long upper lips that wobbled when they suddenly turned their heads.

William had the added distinction of several thin, deeply etched scar marks running down his upper lip from his nose.

Two of the other chimpanzees I got to know well by sight at that time were David Greybeard and Goliath – like David and Goliath in the Bible, these two individuals were closely associated in my mind for they were very often together. Goliath, even in those days of his prime, was not a giant, but he had a splendid physique and the springy movements of an athlete. He probably weighed about one hundred pounds. David Greybeard was less afraid of me, from the start, than were any of the other chimps. I was always pleased when I picked out his handsome face and well marked silvery beard in some chimpanzee group, for, with David to calm the others, I had a better chance of approaching to observe them more closely.

Before the end of my trial period in the field I made two really exciting discoveries – discoveries that made the previous months of frustration well worth while. And for both of them I had David Greybeard to thank.

One day I arrived on the Peak and found a small group of chimps just below me in the upper branches of a thick tree. As I watched I saw that one of them was holding a pink-looking object from which he was, from time to time, pulling pieces with his teeth. There was a female and a youngster and they were both reaching out towards the male, their hands actually touching his mouth. Presently the female picked up a piece of the pink thing and put it to her mouth: it was at this moment that I realised the chimps were eating meat.

After each bite of meat the male picked off some leaves with his lips and chewed them with the flesh. Often, when he had chewed for several minutes on this leafy wadge, he spat out the remains into the waiting hands of the female. Suddenly he dropped a small piece of meat and, like a flash, the youngster swung after it to the ground. But even as he reached to pick it up, the undergrowth exploded and an adult bushpig charged towards him. Screaming, the juvenile leapt back into the tree. The pig remained in the open, snorting and moving backwards and forwards. Presently I made out the shapes of three small striped piglets. Obviously the chimps were eating a baby pig. The size was right, and later, when I realised that the male was David Greybeard, I moved closer and saw that he was indeed eating piglet.

For three hours I watched the chimps feeding. David occasionally let the female bite pieces from the carcass, and once he actually detached a small piece of flesh and placed it in her outstretched hand. When he finally climbed down there was still meat left on the carcass; he carried it away in one hand, followed by the others.

Of course, I was not sure, then, that David Greybeard had caught the pig for himself, but even so, it was tremendously exciting to know that these chimpanzees actually ate meat. Previously scientists had believed that, whilst these apes might occasionally supplement their diet with a few insects or small rodents and the like, they were primarily vegetarians and fruit eaters. No one had suspected that they might hunt larger mammals.

It was within two weeks of this observation that I saw something that excited me even more. By then it was October and the short rains had begun. The blackened slopes were softened by feathery new grass shoots and, in some places, the ground was carpeted by a variety of flowers. The Chimpanzees' Spring, I called it. I had had a frustrating morning, tramping up and down three valleys with never a sign or sound of a chimpanzee. Hauling myself up the steep slope of Mlinda Valley, I headed for the Peak, not only weary but soaking wet from crawling through dense undergrowth. Suddenly I stopped, for I saw a slight movement in the long grass about sixty yards away. Quickly focusing my binoculars, I saw that it was a single chimpanzee – just then he turned in my direction, and I recognised David Greybeard.

Cautiously I moved round so that I could see what he was doing. He was squatting by the red earth mound of a termite nest, and, as I watched, I saw him carefully push a long grass stem down into a hole in the mound. After a moment he withdrew it and picked something from the end with his mouth. I was too far away to make out what he was eating, but it was obvious that he was actually using a grass stem as a tool.

I knew that on two occasions casual observers in West Africa had seen chimpanzees using objects as tools: one had broken open palm nut kernels by using a rock as a hammer, while a group of chimps had been observed pushing sticks into an underground bees' nest and licking off the honey. But somehow I had never dreamed of seeing anything so exciting myself.

For an hour David feasted at the termite mound and then he wandered slowly away. When I was sure he had gone I went over to examine the mound. I found a few crushed insects strewn about, and a swarm of worker termites sealing the entrances of the nest passages into which David had, obviously, been poking his stems. I picked up one of his discarded tools and carefully pushed it into a hole myself. Immediately I felt the pull of several termites as they seized the grass, and when I pulled it out there was a number of worker termites and a few soldiers, with big red heads, clinging on with their mandibles. There they remained, sticking out at right angles to the stem, with their legs waving in the air.

Before I left I trampled down some of the tall dry grass and constructed a rough hide – just a few palm fronds leant up against the low branch of a tree and tied together at the top. I planned to wait there the next day. But it was another week before I was able to watch a chimpanzee 'fishing' for termites again. Twice chimps arrived, but each time they saw me and moved off immediately. Once a swarm of fertile-winged termites – the princes and princesses as they are called – flew off on their nuptial flight, their huge white wings fluttering frantically as they carried the insects higher and higher. Later I realised that it is at this time of year, during the short rains, that the worker termites extend the passages of the nest to the surface, ready for these emigrations. Several such swarms emerge between October and January. It is principally at this time of year that the chimpanzees feed on termites.

On the eighth day of my vigil David Greybeard arrived again, together with Goliath, and the pair worked there for two hours. I could see much better: I observed how they scratched open the sealed-over passage entrances with a thumb or forefinger. I watched how they bit the ends off their tools when they became bent, or used the other end, or discarded them in favour of new ones. Goliath once moved at least fifteen yards from the heap to select a firm-looking piece of vine, and both males often picked three or four stems, whilst they were collecting tools, and put the spares beside them on the ground until they wanted them.

Most exciting of all, on several occasions they picked little leafy twigs and prepared them for use by stripping off the leaves. This was the first recorded example of a wild animal not merely *using* an object as a tool, but actually modifying an object and thus showing the crude beginnings of tool-*making*.

Previously man had been regarded as the only tool-making animal – indeed, one of the clauses commonly accepted in the actual definition of man was that he was a creature who 'made tools to a regular and set pattern'. The chimpanzees, of course, had not made tools to any set pattern: nevertheless, my early observations of their primitive tool-making abilities convinced a number of scientists that it was necessary to redefine man in a more complex manner than before. Or else, as Louis Leakey put it, we should, by definition, have to accept the chimpanzee as Man!

I sent telegrams to Louis about my exciting new observations – the meat-eating and the tool-making – and he was, of course, wildly enthusiastic. Indeed, I believe that the news was helpful to him in his efforts to find further financial support for my work – it was not long afterwards that he wrote to tell me that the National Geographic Society, in the United States, had agreed to grant funds for another year's research.

*Chapter 4* Camp Life

'Memsahib! Memsahib!' Gradually the voice penetrated the depths of my sleep. I sat up. 'Please come. You are wanted,' said the voice, disembodied from behind the light of a small hurricane lamp. It was Adolf. When I asked him what was wrong he was vague, but I gathered it was something to do with a sick baby.

By this time Vanne was awake too, and so we both put on a few clothes and followed Adolf through the darkness. He led us to the little village on the other side of the stream, down on the beach. There the two Game Scouts lived, together with the honorary headman, old Iddi Matata, his large family and about a dozen fishermen in their temporary dry-season huts. Presently we arrived at the large mud-brick and thatched hut of old Iddi. It was past midnight, but everyone was awake, talking and laughing in the smoke-filled main room. Two children scuttled into the shadows as we went in and Iddi's chief wife, who was nursing her twin sons, smiled a greeting to us. Adolf took us to a doorway leading into a smaller room, where it looked very dark, and stood aside for us to go in. And then we realised what it was we were called for. A young woman lay on the earth floor, and beside her a tiny baby still attached to its mother by the umbilical cord. Quite obviously the afterbirth was somehow stuck.

The father was in there, looking rather worried, and one young girl, but no one else seemed to be paying much attention to the situation. We were faced with a dilemma: we wanted to help but knew nothing about midwifery, and if anything happened to the mother we should without doubt be held responsible.

We soon discovered that the baby had been born five hours earlier and that it was a first child. The mother, it seemed, felt only a little pain, but was very cold. We suggested that the cord should be cut and the baby wrapped up, but this was greeted with horrified protest: such conduct would apparently violate time-honoured traditions.

I went for a blanket and some brandy, and roused Dominic to make hot tea, and these things seemed to bring back some life and energy to the poor mother. Vanne and I both felt convinced that some of the other African women there must know far more about childbirth than we did, so we went out to talk to Iddi's head wife with Adolf, who was still hanging around, to translate. She agreed to come and see what she could do as soon as the twins had finished their feed. Presently she came in with a brighter lamp and some warm palm nut oil with which she massaged the girl on her belly and between her thighs, meanwhile gently pulling at the cord. Ten minutes later the afterbirth was delivered. Then, at last, old Iddi came in and, with a pair of ceremonial scissors, proudly cut the cord of his grandchild and himself tied the knot.

We asked Dominic to make the mother some soup, congratulated the father who was beaming with relief, and retired to our beds – having actually done very little but feeling, nevertheless, that we had accomplished quite a lot.

This excursion into midwifery was but one of Vanne's medical tasks for, as custom demanded in those days, we had arrived at the Gombe Stream well supplied with simple medicines – aspirins and ointments, Elastoplast and Epsom Salts. Quite soon after our arrival, Vanne found herself running a well-attended clinic every morning.

David Anstey, before he left, had told the neighbouring Africans that Vanne and I would be pleased to try to treat small ailments, and quite a number had arrived during the first few days – mainly, I suspect, to have a good look at these two strange white women who had so surprisingly abandoned civilisation. But then, one day, a very sick man, with a hugely swollen leg, was helped to our camp. He had two deep tropical ulcers on the lower part of his leg – after Vanne had done a little preliminary cleaning, she realised, to her horror, that the ulceration had already begun to eat away the bone. She begged the man to go to the Kigoma hospital, but he absolutely refused – people only went there to die. And so Vanne treated him herself, with an old-fashioned cure – a saline drip. Every morning and every afternoon the patient sat down with a large bowl of blood-warm salted water which he dripped, very slowly, over his sores. After three weeks the swelling had gone and the wounds were clean. Subsequently it was only a matter of time before he was completely healed.

Word spread fast. After that Vanne's clinics were enormous, and people trudged along the beaches or over the mountains for long distances. One of Rashidi's sons, eight-year-old Jumanne – Swahili for Tuesday – appointed himself as Vanne's orderly and helped her almost every morning, mixing the Epsom Salts, pouring out water for aspirins, cutting the Elastoplast. He was particularly helpful in spotting those who sneaked round to join the queue again, hoping for a second dose! The only payment Jumanne required was a small piece of Elastoplast on some microscopic – or sometimes imaginary – wound.

*     *     *

Vanne's clinics not only cured many maladies but, most important-ly, helped us to establish good relations with our new neighbours. The suspicions which were rife on our arrival were soon dispelled. The Africans, indeed, continued to think we were both slightly crazy, but they were prepared to be friendly for they realised we were genuine. Nor was it long before some of them became in-terested in the work that I was doing. One day Dominic told me that he had heard of an old man, Mbrisho, who had watched four chimpanzees with sticks chasing away a lion. This man lived in a little village up in the hills, just outside the eastern boundary of the Reserve. Would I like to go and talk to Mbrisho and see where the incident had taken place? It seemed an unlikely story, but I knew that lions had occasionally been observed in the area by previous Game Rangers; also I was curious to visit Bubango Village and see the country on the other side of the rift. And so, early one morning, I set off with a man from the village as my guide and the tall willowy Wilbert to interpret, for his English was quite good and my Swahili was still very bad.

It was a long hot climb, and took four hours of pretty steady walking. We passed a procession of African women walking down to the temporary fishing huts on the beach, balancing great bundles on their heads with graceful ease, and chattering and laughing like brightly coloured birds. At one point, when I had stopped to watch a troop of red colobus monkeys, six men passed us, climbing up to

48

I often slept on the Peak
The fishermen attract dagaa to their boats with lights

the village. One of them was old, with a slightly bent back and silver hair, but he seemed quite unconcerned by the steepness of the ascent and the heat of the sun. These men walked with the special springy stride of mountain folk, and each time they placed their stout staffs on to the ground they exhaled their breath with an eerie flute-like whistle.

As we climbed higher the countryside changed. More and more trees were festooned with the feathery grey-green lichen of the upper slopes, and the open spaces were covered with short springy grass so that I was reminded of the Sussex downs. From the top of the rift the view was superb and, in those days, rolling forested country stretched away to the east as far as I could see. To-day much of the forest has been cleared, and African huts and cultivation have marched from all sides towards the boundary of the chimpanzee sanctuary.

The village of Bubango sprawled below us, just on the other side of the mountain-top. It was larger than I had expected. There were groves of banana and palm trees clustered in the greenness of a valley and, on the slopes of the hillside, many patches of cassava, or *muhoge* as it is known locally. Cassava root, ground into a fine white flour and subsequently cooked with water into a porridge, is a staple diet in the area. The huts were mostly rather small and simple, with mud walls and thatched roofs, and a criss-cross of tracks, worn bare by the passage of countless human feet, led from the huts to the stream and to the patches of cultivation. Small children herded goats and sheep, and I even saw a few cows grazing here and there.

The little hut of old Mbrisho stood off to the side of the main dirt track that wound down through the village from the top of the escarpment. He welcomed us in, gave us tea and some delicious doughy African cakes, and beamed widely as he told me the lion story in his slow deep voice. Every so often his conversation was punctuated by a long-drawn out 'Naaaahm.' Mbrisho always uses this word, and I have never been able to discover exactly what it means.

It soon transpired that it was not Mbrisho who had seen the chimps and the lion, but some long dead relative, so that I was none the wiser about that when I left. But I had gained a firm and staunch friend in Mbrisho; he never fails to bring a gift of eggs,

Flo using a grass tool to catch termites
*Copyright National Geographic Society*

carefully wrapped in a piece of cloth, when he comes down the mountain to visit our camp. And, when a man lives off the land, and is old, a few eggs are a valuable present and one which we are proud to accept.

Mbrisho, like most of the able-bodied men of the surrounding villages, was a fisherman until he 'retired.' The principal catch is the small, sardine-sized *dagaa* which is caught at night in orange- or red-dyed nets shaped like gigantic butterfly nets. Each canoe, manned by two fishermen, is equipped with two or more bright paraffin pressure lamps: the fish, attracted by the light, rise to the surface and are scooped up in the net. Whenever a big shoal is sighted the fishermen start to sing, stamping with their feet, banging against the boat with paddle or net handle and this, apparently, encourages the fish to rise. When fishing is good it sounds as though all the human inhabitants for miles around are having a fantastic party on the lake.

When the bottom of a canoe is filled with fish the catch is paddled back to the shore, where other men start spreading it out on special drying beds of fine gravel. If fishing is good, each canoe may return two or three times in a night, and the rising sun is reflected from millions and millions of tiny scales so that the beaches seem changed to silver.

During the day the fishermen, or their wives or children, periodically work their way up and down the drying beds, prodding at the fish with long sticks, turning them so that the sun dries them evenly on both sides. And then in the evening the catch is gathered up into sacks whilst the men who will go out fishing sit around their grass huts talking, and their women folk prepare the evening meal of *ugali*, or cassava porridge, mixed with dried *dagaa* that have been fried in red palm nut oil.

When the moon is bright the *dagaa* are no longer attracted towards the lamps of the canoes, and during this period the dried fish are taken to Kigoma for sale. Mostly they are loaded on to the small thirty-foot motorised boats which ply up and down the lake shore. At the peak of the season these water taxis, as we call them, are loaded with sack upon sack of fish, for the *dagaa* industry is a profitable one. A great deal is sold locally, but even more is railed away to other parts of East Africa and to the great copper mines of Nyanza in the south.

The fishermen who do not go to Kigoma return to their home villages to visit their families, and so, for ten days or so each month, the beaches of the Gombe Stream are deserted. At such times I loved walking along the lake-shore, when my work took me to some distant valley. In the early morning I occasionally met the lumbering shape of a hippo returning to the water after a night's feeding on the lush grass along the lake-shore. Bushbucks and bushpigs often moved along the beaches between the valleys, and once I saw a small herd of buffalo, looking huge and very black against the white sand. There was always a chance, too, of seeing some of the smaller animals – one of the mongooses, perhaps, or a slender dainty genet with its ringed tail, or a larger thick-set civet cat.

One evening when I was wading in the shallows of the lake to pass a rocky outcrop, I suddenly stopped dead as I saw the sinuous black body of a snake in the water. It was all of six feet long, and from the slight hood and the dark stripes at the back of the neck, I knew it to be a Storm's water cobra – a deadly reptile for the bite of which there was, at that time, no serum. As I stared at it, an incoming wave gently deposited part of its body on one of my feet. I remained motionless, not even breathing, until the wave rolled back into the lake, drawing the snake with it. Then I leapt out of the water as fast as I could, my heart hammering.

A few weeks before I had encountered another cobra – one of the white-lipped variety which, aiming for its victim's eyes, can spit venom from as far as six feet and cause temporary, or even permanent, blindness. At the time I had been standing looking over the valley and had glanced down to see the snake gliding between my legs. It had paused momentarily, testing the canvas of my shoe with its sensitive flickering tongue. But I felt no fear whatsoever for there had been no chance of a sudden wave dashing it against me and making it frightened, no suction of water that might coil it around my ankle. There is something utterly unnerving about a snake in the water.

Lake Tanganyika is one of the few areas of fresh water in East Africa which is virtually free of the dreaded bilharzia snail, at least in the Kigoma and Gombe area, and the water is cool, sparkling clean and excellent for swimming. In those early days I never had time for swimming – not that I felt any great desire to plunge into

the water after my experience with the snake – and after hearing
about Chiko's experience. Chiko was Dominic's wife who, with her
little girl, had joined us soon after our arrival at Gombe. One day
when she was standing in the shallow water doing her washing,
there was a sudden swirling a few feet away and, with a shriek she
leapt out on to the shore, staring in horror at the place where,
but a moment before, she had seen a crocodile. It wasn't a very big
crocodile – I had myself seen it a few times swimming in the lake –
but I should not have liked to meet it in the water. Chiko's experience, however, became the joke of the moment – every African I
met for the next few weeks was laughing about the crocodile that
tried to grab my cook's wife. Dominic himself, when first he told
me, laughed till the tears ran down his cheeks.

\*     \*     \*

It is when the fishermen leave their huts during the full moon
that the baboons come into their own on the beaches. All along
the shore-line, where there are drying beds, the troops gather,
searching through the pebbles for left-over scraps of dried fish.
They sift through the sand around the huts too, probably looking
for specks of cassava scattered where the women pounded the roots
into flour for porridge. The Africans remove anything of value
when they leave the beaches, for the baboons are very destructive.
I have watched them pulling apart a whole roof as they search
through the thatch for insects, and they move in and out of the
huts as though they own them, eating anything edible, investi-
gating or pulling apart anything that is not.

The baboons very soon made themselves at home around our
camp too, and Vanne quickly learned never to leave the tents un-
guarded. About two weeks after our arrival she went for a short
walk: when she returned it was to find our belongings strewn in
all directions, and one blasé male baboon sitting by the overturned
table polishing off the loaf that Dominic had baked that morning.
Vanne was particularly incensed by the way the others barked at
her, from the surrounding trees, as though challenging her right
to her own home. It was not long after this episode that Vanne,
during a visit to the camp loo, or *choo* as it is known in East Africa,
looked up to see five large old male baboons sitting round in a

semi-circle and watching her. She told me that evening that she had felt frightfully embarrassed for a moment.

It was far worse when, one morning, Vanne, who had been dozing after my early departure, suddenly heard a small sound in the tent. She opened her eyes and there, silhouetted in the entrance, she saw a huge male baboon. He and she remained motionless for a few moments and then he opened his mouth in a huge yawn of threat. In the grey light of dawn Vanne could just see the gleam of his teeth and she thought her last hour had come. With a sudden yell she sat bolt upright in bed, waving her arms – and her unwelcome visitor fled. He was a horrible baboon, that one – an old male who took to hanging around our camp at all hours of the day, lurking in the undergrowth and dashing out whenever opportunity presented to steal a loaf of bread or some other item of food. We called him Shaitani, which is Swahili for devil, and we were immensely relieved when he suddenly disappeared.

In those days food was precious; not only because of our limited budget, but also because both Vanne and I hated 'Kigoma days' – when we had to go into the town to renew our supplies and collect our mail. We went as little as possible, but once every three or four weeks we had to make the expedition. We used to set off at about six in the morning when the lake was usually calm, and breakfast at the hotel when we arrived. Then we trailed round doing the shopping, bargaining at the fruit and vegetable market, ordering tinned goods for the next month, and queueing in the post office. We always had a pleasant interlude at midday, for we were invariably invited to lunch by one or other of the friends we had made during our initial stay in Kigoma. Indeed, they often tried to persuade us to stay the night and most of them thought I was merely being anti-social when I explained that it was bad enough to leave my work for one day and that I couldn't think of staying away for a second.

At first we used to take Dominic with us on these expeditions. He was amazingly loyal and would bargain heatedly in the market so that we should not be overcharged by as much as a farthing. But after a while we realised that it was better to leave Dominic to look after camp. The locally made beer, brewed from fermented bananas, is strong – and drinking is Dominic's weakness. On several occasions we could not find him at all when it came to the appointed time to leave – once he did not turn up again for a week, and we had to manage without him. And we shall never forget the day when, though we did manage to find him in time to return with us, he was in a very merry state. Laughing at nothing, he accompanied us to the boat and half-stepped, half-fell in. He is an intelligent man, and often very amusing when a little drunk. He soon had Vanne and I laughing with him, and we were a gay party as we cast off and chugged out of Kigoma bay.

By this time it was almost dark, for the search for Dominic had delayed us. Normally I steered a course quite close to shore, but that evening the water was dotted with fishermen's canoes, paddling out from the shore, and they were hard to see in the dusk. So I drove the boat along half a mile or more from the beach. Quite suddenly, when we were about quarter of the way home, the engine stopped. Nothing we did produced even a spark, and none of us knew anything about engines. All we could do was to row to shore.

Dominic announced proudly that he would row us in. Seating himself centrally he seized the two oars, settled himself, and gave a mighty pull. He caught a whopping crab and the next instant was lying flat out on top of one of the large baskets of fruit from the market. He could not get out for laughing. Eventually we hauled him back to the seat, and I said I would row. But this was too much for his sense of propriety – at least he must take one oar. For the next ten minutes we proceeded to turn small circles around his oar. Finally I persuaded him that I could manage, that I loved rowing more than anything, and off we set.

When we eventually reached the beach we found, to our delight, that one of the water taxis was anchored there. A crowd of fishermen soon gathered around us and before long the driver of the taxi was found. After much arguing he agreed to transport us and our boat back to camp. But Dominic was incensed at the price we were to be charged. Telling us to wait, he vanished into the night before we could stop him.

We had a long wait, but we didn't want to leave without Dominic: also we were grateful to him for thinking of our purse for it was, indeed, very slender. Thirty minutes later he returned with four stalwart Africans. These men, Dominic informed us, would *paddle* us home. They were friends of his, and they would only charge us a quarter of the price demanded by the motorised water taxi. This sounded marvellous until we discovered that even with eight men rowing hard it would take more than eight hours to get home! I think Dominic never quite forgave us for taking the faster vessel and allowing ourselves to be robbed.

It was not long after this that Vanne, who had gone to Kigoma without me, had a much worse experience. Dominic was not with her, but the willowy Wilbert had asked to go in as he wanted to do some shopping. He and Vanne parted in the morning, and arranged to meet back at the boat at five o'clock. When he finally appeared, half an hour late, Vanne's heart sank: here was another victim of Kigoma's banana beer! Wilbert, staggering slightly, swayed up to the boat and then, to Vanne's horror, pulled out a knife which he waved around whilst, with staring eyes, he began to mutter incoherently about death and revenge.

Vanne admitted that she had been frightened, for Wilbert, willowy though he appeared, was a very large man indeed, and he looked terrifying with his eyes bloodshot and the blade of the knife glinting in the evening sun. However, she kept her head, spoke to him quietly, and asked him to let her look after the knife in case he cut himself getting into the boat. To her surprise he stared hard at her and then a bewildered expression came over his face. He lurched towards her, handed her the knife and without another word, climbed into the boat. He was very quiet during the trip home. We never did discover what it was all about, and Wilbert never asked for his knife back again.

How lucky I was to have a mother like Vanne – a mother in a million. I could not have done without her during those early days. She ran the clinic and ensured the goodwill of my neighbours, she kept the camp neat, she pressed and dried for me the specimens of the chimps' food plants that I collected and, above all, she helped me to keep up my spirits during the depressing weeks when I could get nowhere near the chimps. How nice it was to come back in the evening and find a warm welcome. How pleasant to be able to discuss the events of the day, frustrating or exciting, over the fire during supper, and to hear the gossip of camp.

Vanne put up with the most primitive conditions without a murmur. We lived, in those early days, on baked beans, corned beef and other tinned meats and vegetables – for we had no refrigerator.

We bathed, at night, in a tiny canvas 'tub' supported at each corner by a wooden frame and filled with a few inches of water – too hot when one got in and cold when one got out a few moments later. Sometimes giant hairy spiders took refuge in our tent, and twice Vanne woke in the morning to see the flattened, evil-looking body of a deadly giant centipede clinging to the roof of the tent above her bed. And, as if all this wasn't enough, there was something in the water that disagreed with Vanne's never strong stomach so that she seldom, if ever, felt one hundred per cent fit.

Five months after our arrival Vanne had to return to England. The authorities in Kigoma no longer worried about me being on my own; I was, by then, part of the Gombe Stream landscape and, thanks to Vanne's clinics, I enjoyed excellent relationships with the local inhabitants. Just before Vanne left we were joined by Hassan, our old friend from Lake Victoria. We were overjoyed to see him, and Vanne was able to go knowing that I was in reliable hands. Hassan would drive the little boat and do the awful 'Kigoma days' on his own, and if anything happened, Hassan, somehow, would cope. For Hassan had often proved himself in times of crisis during his thirty years with Louis.

My fire seemed lonely when Vanne had gone. Even Terry, the toad, who had become our nightly companion, served only to remind me of the times Vanne and I had laughed at him together as he gorged himself on the insects around the lamp. And on the evenings when Crescent, the genet, came softly towards the tent to take a banana, I found myself wishing Vanne was there to share in my appreciation of her slender grace.

As the weeks passed, however, I accepted aloneness as a way of life, and I was no longer lonely. I was utterly absorbed in my work, fascinated by the chimps, too busy in the evenings to brood. In fact, had I been alone for longer than a year I might have become a rather strange person, for inanimate objects began to develop their own identities: I found myself saying 'Good morning' to my little hut on the Peak; 'Hallo' to the stream where I collected my water. And I became immensely aware of trees – just to feel the roughness of a gnarled trunk or the cold smoothness of young bark with my hand, filled me with a strange knowledge of the roots under the ground and the pulsing sap within. I longed to be able to swing through the branches like the chimps, to sleep in the tree-tops, lulled by the rustling of the leaves in the breeze. In particular, I loved to sit in a forest when it was raining, and to hear the pattering of the drops on the leaves and feel utterly enclosed in a dim twilight world of greens and browns and dampness.

# Chapter 5 The Rains

Soon after Vanne's departure the light rains of the Chimpanzees' Spring gave place to the long rains. Showers became deluges which sometimes lasted with unabated fury for two hours or more. One of those wild storms, which occurred a week or so after the change of seasons, I shall long remember. For two hours I had been watching a group of chimpanzees feeding in a huge fig tree. It had been grey and overcast all morning, with thunder growling in the distance.

At about noon the first heavy drops of rain began to fall. The chimpanzees climbed out of the tree and, one after the other, plodded up the steep grassy slope towards the open ridge at the top. There were seven adult males in the group, including Goliath and David Greybeard, several females, and a few youngsters. As they reached the ridge the chimpanzees paused. At that moment the storm broke. The rain was torrential and the sudden clap of thunder, right overhead, made me jump. As if this was a signal, one of the big males stood upright and, as he swayed and swaggered rhythmically from foot to foot, I could just hear the rising crescendo of his pant-hoots above the beating of the rain. Then he charged off, flat-out down the slope towards the trees he had just left. He ran some thirty yards and then, swinging round the trunk of a small tree to break his headlong rush, leapt into the low branches and sat motionless.

Almost at once two other males charged after him. One of them broke off a branch from a tree as he ran and brandished it in the air before hurling it ahead of him. The other, as he reached the end of his run, stood upright and rhythmically swayed the branches of a tree back and forth, before seizing a huge bough and dragging it farther down the slope. A fourth male, as he too charged, leapt into a tree and, almost without breaking his speed, tore off a large branch, leapt with it to the ground and continued down the slope.

As the last two males called and charged down, so the one who had started the whole performance climbed from his tree and began plodding up the slope again. The others, who had also climbed into trees near the bottom of the slope, followed suit. And then, when they reached the ridge, they started charging down all over again, one after the other, with equal vigour.

The females and youngsters had climbed into trees near the top of the rise as soon as the displays had begun, and there they remained watching throughout the whole performance. As the males charged down and plodded back up, so the rain fell harder and harder, jagged forks or brilliant flares of lightning lit the leaden sky and the crashing of the thunder seemed to shake the very mountains.

My enthusiasm was not merely scientific as I watched, enthralled, from my grandstand seat on the opposite side of the narrow ravine, sheltering under a sheet of polythene. In fact, it was raining and blowing far too hard for me to get at my notebook or use my binoculars: I could only watch, and marvel at the magnificence of those splendid creatures. With a display of strength and vigour such as this, primitive man himself might have challenged the elements.

Twenty minutes from the start of the performance the last of the males plodded back up the slope for the last time. The females and youngsters climbed down from their trees and the whole group moved over the crest of the ridge. One male paused and, with his hand on a tree-trunk, looked back – the actor taking his final curtain. Then he too vanished over the ridge.

I continued to sit for a while, staring almost in disbelief at the white scars on the tree-trunks and the discarded branches on the grass – all that remained, in that rain-lashed landscape, to prove that the wild 'rain dance' had taken place at all.

I should have been even more amazed had I known then that I should only see such a display twice more in the next ten years. Often, indeed, male chimpanzees react to the start of heavy rain by performing a rain dance, but this is usually an individual affair. Yet I have only to close my eyes to see again, in vivid detail, that first spectacular performance.

\*       \*       \*

As the rainy season progressed the grass shot up until it was over twelve feet in some places, and even on the exposed ridges it often reached heights of at least six feet. When I left the tracks which I knew – if, indeed, I could find them at all – I could not tell where I was going, and had to stop every so often and climb a tree to get my bearings. Also I was no longer able to sit down, wherever I happened to be or wherever was convenient, when I came across a group of chimpanzees, for usually my view would then be totally obscured by grasses. I have never been able to work with binoculars for long periods of time when standing, so I had either to bend down hundreds of grass stems or else climb a tree. As the rainy season progressed I became increasingly arboreal in my habits, but despite my love for trees it was not very satisfactory for I lost time both in looking for a suitable tree and in breaking away branches that obstructed my view of the chimps. And when there was a wind, which was often, I couldn't keep the binoculars steady anyway.

I found it difficult, also, to shield my binoculars from the rain. I made a sort of polythene tube which kept out much of the wet, and I pulled a large piece of polythene forward over my head, like a peaked cap, when I was watching chimps. Even so there were many days when I couldn't use my binoculars because they were clouded over inside with droplets of condensed moisture.

Even when it was not actually raining, the long grass remained drenched nearly all day, either with rain or the heavy nightly dew, and there were periods when I seemed to be wet through for days on end. Indeed, I think I spent some of the coldest hours of my life in those mountains, sitting in clammy clothes in an icy wind watching chimpanzees. There was even a time when I dreaded the early morning climb to the Peak: I left my warm bed in the darkness, had my slice of bread and cup of coffee by the cosy glow of a hurricane lamp, and then had to steel myself for the plunge into that icy, water-drenched grass. After a while, though, I took to bundling my clothes into a polythene bag and carrying them: there was no one to see my ascent and it was dark anyway. Then, when I knew there were dry clothes to put on when I reached my destination, the shock of the cold grass against my naked skin was a sensual pleasure. For the first few days my body was criss-crossed by scratches from the tooth-edged grass, but after that my skin hardened.

One morning, in the first grey light of dawn, I plodded up the last steep slope to the Peak. As I reached the top, my foot seemed to freeze in mid-air, for there, no less than four yards from me, a lone buffalo lay, half hidden in the long grass. He must have been asleep, or surely he would have heard me; and luckily the breeze was blowing his rich, cow-like scent directly to my nostrils. Had it been the other way around ... As it was, I was able to creep quietly away, and he was not disturbed.

It was at this time too that a leopard actually passed within a few yards of me as I sat in some long grass. I never knew he was there until I saw the white tip of his tail just ahead of me, and there was no time to retreat. So I just held my breath. I doubt if he ever knew how close to a human he had been!

On the whole I loved that rainy season at the Gombe. It was cool most of the time and there was no heat to distort my long-distance observations. I have always loved to feel myself an integral part of nature, and the crunching of my feet on the crackling leaf carpet of the forest floor in the dry season had always bothered me. When the leaves became soft and damp during the rains I could move through the trees silently, and catch more than fleeting glimpses of some of the shyer creatures. Best of all, of course, I was continually learning more and more about chimpanzees and their behaviour.

In the dry season the chimpanzees, I knew, usually rested on the ground at midday for I had often glimpsed them lying stretched out in the shade. In the rainy season though, the ground is frequently sodden and I found the chimpanzees made quite elaborate day nests on which to rest. To my surprise they often made these whilst it was still actually raining and then sat, hunched up with their arms round their knees and their heads bowed, until the rain stopped. In the mornings they got up later too and quite often after feeding or just sitting about for two or three hours, made new nests and lay down again. I suppose this was because they sometimes had such miserably wet and cold nights that they couldn't sleep, and so were tired in the morning. They usually went to bed earlier too. Sometimes, when I left the chimps in their nests, soaking wet from a late afternoon storm, I felt not only desperately sorry for them, but guilty because I was returning to a hot meal, dry clothes and a tent. And I felt even worse when I woke in the middle of the night to hear the rain lashing down on to the canvas and thought of all the poor huddled chimps shivering on their leafy platforms whilst I snuggled cosily into my warm bed.

Sometimes, at the beginning of a storm, a chimpanzee would shelter under an overhanging trunk or tangle of vegetation, but then, when the rain began to drip through, he usually emerged and just sat, hunched and looking miserable, in the open. Small infants seemed to fare the best in a heavy storm. Quite often I sàw old Flo who, of all the females was least afraid of me at that time, sitting hunched over two-year-old Fifi. At the end of a deluge Fifi would crawl from her mother's embrace looking completely dry. Flo's son Figan, about four years older than Fifi, often swung wildly through the tree on such occasions, dangling from one hand and kicking his legs, leaping from branch to branch, jumping up and down above old Flo until she was showered with debris and hunched even lower to avoid the twigs that lashed her face. It was a good way of keeping his blood warm – rather like the wild rain display with which older males often greeted the start of heavy rain.

As the weeks went by I found that I could usually get closer to a group of chimpanzees when it was cold and wet than when the weather was dry. It was as though they were too fed-up with the conditions to bother about me. One day I was moving silently through the dripping forest. Overhead the rain pattered on to the leaves, and all around it dripped from leaf to leaf to the ground. The smell of rotten wood and wet vegetation was pungent: under my hands the tree-trunks were cold and slippery and alive. I could feel the water tickling through my hair and running warmly into my neck. I was looking for a group of chimps I had heard before the rain began.

Suddenly, only a few yards ahead of me, I saw a black shape hunched up on the ground with its back to me. Quickly I hunched down on to the ground myself: the chimp hadn't seen me. For a few minutes there was silence save for the pattering of the rain, and then I heard a slight rustle and a soft *hoo* to my right. Slowly I turned my head, but I saw nothing in the thick undergrowth.

When I looked back, the dark shape that had been in front of me had vanished. Then came a sound from above: I looked up and there saw a large male directly overhead: it was Goliath. He stared down at me with his lips tensed and very slightly shook a branch. I looked away, for a prolonged stare can be interpreted as a threat. I heard another rustle to my left, and when I looked I could just make out the black shape of a chimp behind a tangle of vines. When I looked ahead I saw two eyes staring towards me and a large hand gripping a hanging liana. Another soft *hoo*, this time from behind. I was surrounded.

All at once Goliath uttered a long drawn-out *wraaaa* and I was showered with rain and twigs as he threatened me, shaking the branches. The call was taken up by the other dimly-seen chimps. To me it is one of the most savage sounds of the African forests, second only to the trumpeting scream of an enraged elephant. All my instincts bade me flee, yet I forced myself to stay there, trying to appear disinterested and pretending to eat roots from the ground. The end of the branch above me hit my head: with a stamping and slapping of the ground a black shape charged through the undergrowth ahead, veering away from me at the last minute and running at a tangent into the forest. I think I expected to be torn to pieces: I do not know how long I crouched there, tensed and horrified, before I realised that everything was still and silent again, save for the drip drip of the rain-drops. Cautiously I looked round. The black hand and the glaring eyes were no longer there; the branch where Goliath had been was deserted; all the chimpanzees had gone. Admittedly my knees shook a little when I got up, but there was that sense of exhilaration that comes when danger has threatened and left one unharmed – and the chimpanzees were surely less afraid of me!

It lasted for about five months, this period of aggression and hostility towards me, following the initial fear and hasty retreat that had taken place whenever the chimps saw me. There is one other incident which stands out in my memory: it took place about three weeks after the one I have just described. I was waiting, on one side of a narrow ravine, hoping that chimpanzees would arrive in a fruit-laden tree on the opposite slope. When I heard the deliberate footsteps of approaching chimpanzees behind me I lay down flat and kept still – for often, if the apes saw me on their way to feed, they changed their minds and fed elsewhere. Once they had started to feed, however, once they were surrounded by the delicious taste and sight and smell of the fruit, their hunger usually proved stronger than their distrust of me. On this occasion the footsteps came on and suddenly stopped, quite close. There was a soft *hoo*, the call of a worried, slightly fearful individual. I had been seen. I kept quite still and presently the footsteps came even closer. Then I heard one chimpanzee suddenly run for a few yards; this was followed by a loud scream. What on earth, I wondered, was happening?

A moment later I saw a male chimpanzee climbing a tree only a couple of yards away. He moved over into the branches above my head and began screaming at me, short, loud, high-pitched sounds, with his mouth open. I stared up into his dark face and brown eyes. He began climbing down towards me until he was no more than ten feet above me and I could see his yellow teeth and, right inside his mouth, the pinkness of his tongue. He shook a branch, showering me with twigs. Next he hit the trunk and shook more branches and, all the while, continued to scream and scream and work himself into a frenzy of rage. Suddenly he climbed down and went out of sight behind me.

It was then that I saw a female with a tiny baby and an older child sitting in another tree and staring at me with wide eyes. They were quite silent and quite still. I could hear the old male moving about behind me and then his footsteps stopped. He was so close I could hear his breathing. It seemed unreal.

All at once there was a loud bark, a stamping in the leaves, and my head was hit, hard. At this I had to move, had to sit up, had to believe it was all really happening. The male was standing looking at me and for a moment I believed he would charge. But he turned and moved off, stopping often to turn and stare at me. The female and youngster climbed down silently and moved after him. A few moments later I was alone again, my heart beating fast. But I felt a sense of triumph; I had made real contact with a wild chimpanzee – or perhaps it should be the other way round.

When I looked back, some years later, at my description of that male, I am certain that it was the bad-tempered, irascible, paunchy J.B. The behaviour fits in perfectly with the irritable, fearless character that I later came to know so well. I suppose he was

Weak from fever I struggled up the mountain and found the Peak

Young chimpanzees are very agile

Chimpanzees make nests to sleep in

Some chimpanzees have a range of about thirty square miles

David, Goliath and William raided my tent for cardboard to chew on

puzzled by my immobility and the sheet of polythene that was protecting me from the light rain. He simply had to find out exactly what I was and make me move – he must have known, from my eyes, that I was alive.

It was after incidents of this sort that I longed to return to camp and find Vanne there, so that I could share with her the horror or the delight of the occasion. But I told Dominic and Hassan about my encounter with the old male – and they in turn told old Iddi Matata. He came to see me the following evening and recounted the story of an African who had climbed half-way up a palm tree to collect the ripe fruit without noticing that a male chimp was feeding at the top. The ape suddenly saw him and rushed down, hitting out at the African's face as he passed. The man lost one eye. And so the rumour went around that I had some magic about me, that I went unharmed where others were hurt, that I was no ordinary English girl. It all helped my friendly relations with my African neighbours.

The long rains should end during April, I had been told. But that year it was still raining in June, although less frequently. In between the cold grey days the whole area was like a gigantic tropical green-house. The moisture that was sucked by the sun from the lush vegetation of the mountains was trapped in the valleys, trapped between the long grass stems. Climbing the steep slopes was often a nightmare. Sometimes I felt I simply had to climb into a tree in order to breathe, and once I was up there I wondered why on earth our ancestors had ever left the branches. I think, when I look back over that May and June, that they were the worst of all my early months – worse even than the time when the chimpanzees fled at my approach. I had several returns of fever: often it was an effort to lift a hand, so bad was the humidity, let alone struggle up mountains. And the chimpanzees, who had been feeding in large noisy groups, were splitting up more and more often into small units of two to six. Frequently such groups made no sounds all day as they wandered through the forests feeding on the fruit of the ubiquitous mbula tree known as wild custard apples.

Gradually, however, the terrible humidity lessened, strong winds blew daily down through the valleys and my spirits and health were restored. The wild fig season came round again and this time, instead of watching from the Peak, I was able to move down into the valley and sit quite close to the trees where the chimpanzees fed each day.

Once, as I was watching a group in a tree about thirty yards away, I heard a slight rustle in the leaves behind me. I looked round and there, about fifteen feet away, sat a chimpanzee with his back to me. I kept quite still, thinking he had not seen me, but after a few moments he glanced casually at me over his shoulder, then went on chewing. He stayed there for another ten minutes, sometimes giving me a quick look, before finally walking away. That was Mike, an adult male with a face almost as handsome as that of David Greybeard. The incident took place a few weeks after that never to be forgotten day, described at the beginning of the book, when David Greybeard and Goliath also sat calmly only a few feet away from me. Their original fear of me had gradually given place to aggression and hostility, and now many of the chimps had begun to accept me as part of their normal, everyday landscape. A strange white ape, very unusual to be sure, but not, after all, terribly alarming.

Towards the end of August my sister Judy arrived from England. The National Geographic Society, which was by then financing my research, was, of course, interested in obtaining photographs for its magazine. The Society wanted to send a professional photographer, but I was terrified at the thought of a stranger arriving on the scene and ruining my hard-won relationship with the chimps. I suggested to Louis that Judy should come out – not because she had had any photographic experience, but because she looked like me and would be willing to sacrifice chances of getting pictures for the good of my work. The National Geographic Society did not finance her trip, but a British week-end newspaper, *Reveille*, took a gamble and paid all her expenses in return for a series of interviews with me when I returned to England.

Poor Judy. When she arrived it was virtually the end of the dry season – a dry season that had lasted no more than six weeks. I had built small hides near trees which I expected to fruit during September and October, but the crops were poor that year – and it rained nearly every day. Judy spent hours and hours waiting in hides in the pouring rain, trying to shelter herself and her photographic paraphernalia under a sheet of polythene. The chimps hardly ever came and, when they did, it was raining too hard for

Judy to take a single picture. However, during November things cheered up a little and she was able to take some of the first-ever pictures of chimpanzees using tools as they fished for termites. Also she got some shots of me and camp life and the fishermen, which made her trip worth while so far as her sponsors were concerned.

When Judy first arrived at the Gombe Stream she was horrified at my emaciated appearance. For eighteen months, apart from the occasional 'Kigoma day,' or short periods when fever had laid me low, I had worked full-time in the mountains. My alarm clock was always set for 5.30 in the morning and, after a slice of bread and a cup of coffee, I hurried off after my chimps. I never felt the need for food when I was roaming the forests: indeed, I was lucky enough to require no water either, except on rare occasions. The cups of coffee I made on the Peak were luxuries. And then, after returning to camp as darkness fell, there were always my notes to transcribe – often I was working until far into the night. No wonder I had lost weight.

Judy felt it her duty to 'fatten me up.' Accordingly she ordered such things as porridge and custard, Bovril and Horlicks. But somehow I never felt like eating them. And, since Judy couldn't bear to see them wasted, she ate them instead.

In December we had to pack up camp – in the pouring rain, of course – and store all my equipment in Kigoma. For Louis Leakey had managed to get me admitted to Cambridge University to work for a PhD degree in ethology, the study of animal behaviour. I was to be one of the few students admitted as a PhD student without ever having sat for a B.A. degree. Louis met Judy and me in Nairobi and sent a cable to Vanne. It read: 'GIRLS ARRIVED SAFELY STOP ONE THIN ONE FAT.'

*Chapter 6*  The Chimps Come to Camp

1961 was a cold winter in England, and in Cambridge, with the winds whistling over the flat country straight from the icy wastes of Norway, the snow and the frost and the frozen water pipes seemed interminable. I felt utterly remote from Africa and the chimpanzees and all that I longed for most at that time. Of course, I was immensely grateful for the privilege of going to Cambridge, and for the chance of working under the supervision of Professor Robert Hinde – but what was David Greybeard doing in the meantime? How were Goliath and Flo? What was I missing?

At long last the spring thawed the hard ground; in two months I would be back in Africa. First I had two ordeals to face, and the thought of them terrified me far more than any encounter with enraged chimpanzees. I had to speak at international conferences, in London and in New York, for other scientists were avid for first-hand information about the ways of my chimpanzees. But these milestones came and went and at last, unbelievably, the six months' exile were over and I was heading for Africa again, crossing the vastness of the Sahara Desert in the lurid red light of dawn that is so much an integral part of modern air travel.

Then the fears that had haunted me in England came to a head. Would the chimpanzees have forgotten me? Would I have to get them used to me all over again? I need not have worried; when I got back to the Gombe Stream it seemed that the chimps were, if anything, *more* tolerant of my presence than before. I resumed my work in the mountains as though I had never been away.

One evening I returned to camp and found Dominic and Hassan very excited. A large male chimpanzee, they told me, had walked right into camp and spent an hour feeding from the palm tree that shaded my tent. The following evening I learned that the same chimp had paid another visit. I determined to stay down the next day to see if he came again.

It seemed strange to lie in my bed and watch the dawn break, to have breakfast in camp, to sit in my tent in the *daylight* to type out my previous day's notes on the typewriter I had brought back from England. And it seemed quite unbelievable when, at about ten o'clock, David Greybeard strolled calmly past the front of my tent and climbed the palm tree. I peeped out and heard him giving low-pitched grunts of pleasure as he poked the first red fruit from its horny case. An hour later he climbed down, paused to look, quite deliberately, into the tent, and wandered off. After all those months of despair, when the chimpanzees had fled at the mere sight of me five hundred yards away, here was one making himself at home in my very camp. No wonder I found it hard to believe.

David Greybeard paid regular visits until the palm tree's fruit was finished, and then he stopped coming. These oil nut palms do not all fruit at the same time, however, and so, a few weeks later, the hard nut-like fruits on another of the camp palms became red and ripe. David Greybeard resumed his daily visits. I did not often stay to watch him for there was a limit to the amount of information I could gain from watching a lone male guzzling palm nuts – sometimes, though, I waited for him to come just for the intense pleasure of seeing him so close and so unafraid. One day, as I sat on the veranda of the tent, David climbed down from his tree and then, in his deliberate way, walked straight towards me. When he was about five feet from me he stopped and, slowly, his hair began to stand on end until he looked enormous and very fierce. A chimpanzee may erect his hair when he is angry, frustrated or nervous: why, I wondered somewhat apprehensively, had David put his hair out? All at once he ran straight at me, snatched up a banana from my table, and hurried off to eat it farther away. Gradually his hair returned to its normal sleeked position.

After that incident I asked Dominic to leave bananas out whenever he saw David, and so, even when there were no ripe palm nuts, the chimp still wandered into camp sometimes, looking for bananas. But his visits were irregular and unpredictable, and I no longer waited in my tent for him to come.

About eight weeks after my return to the Gombe I had a slight attack of malaria. I stayed in bed and, hoping that David Greybeard might pass by, asked Dominic to set out some bananas. Late that morning David walked up to my tent and helped himself to the fruit. As he walked back towards the bushes I suddenly saw that a second chimpanzee was standing there, half hidden in the vegetation. Quickly I reached for my binoculars – it was Goliath. As David sat down to eat, Goliath went close to him, peering into his face. As David rolled a banana skin wadge around in his lower lip, squeezing out the last juices and occasionally pushing it forward to look down at it over his nose, Goliath reached out one hand to his friend's mouth, begging for the wadge. Presently David responded by spitting out the well-chewed mass into Goliath's hand, and then Goliath sucked on it in turn.

The next day Goliath followed David to camp again. I kept hidden right inside my tent, with the flap down, and peeped out through a small hole. This time Goliath, with much hesitation and with all his hair on end, followed David to the tent and seized some bananas for himself.

The following few weeks were momentous. I sent Hassan off to the village of Mwamgongo, to the north of the Reserve, to buy a supply of bananas, and each day I put out a pile near my tent. The figs in the home valley were ripe again so that large groups of chimps were constantly passing near camp. I spent part of my time up near the fig trees, and the remainder waiting for David in camp. He came, nearly every day, and not only did Goliath sometimes accompany him, but William also, and, occasionally, a youngster.

One day, when David came by himself, I held a banana out to him in my hand. He approached, put his hair out and gave a quick soft exhalation, rather like a cough, whilst at the same time he jerked his chin up. He was mildly threatening me. Suddenly he stood upright, swaggered slightly from foot to foot, slapped the trunk of a palm tree beside him with one hand, and then, very gently, took the banana from me.

The first time I offered Goliath a banana from my hand was very different. He, too, put his hair out, then seized a chair and charged past me, almost knocking me over. Then he sat and glowered from the bushes. It was a long time before he behaved as calmly in my presence as David; if I made a sudden movement and startled him he often threatened me quite vigorously, uttering a soft barking sound whilst rapidly raising one arm or swaying branches with quick rather jerky movements.

It was exciting to be able to make fairly regular observations on the same individuals, for previously I had found this almost impossible. Chimpanzees follow no set route on their daily wanderings and so, except when a good fruit tree had been ripe, it had only been by chance that I had observed the same chimpanzee more than a few times in any one month. In camp, however, I was able to make detailed observations on social interactions between David, Goliath and William several times a week. Often, too, I saw incidents involving one of the three when I was watching the chimpanzee groups feeding on figs in the home valley.

It was at this point that I began to suspect that Goliath might be the highest-ranking male chimpanzee in the area – and later I found that this was indeed the case. If William and Goliath started to move towards the same banana at the same time, it was William who gave way and Goliath who took the fruit. If Goliath met another adult male along a narrow forest track he continued – the other stepped aside. Goliath was nearly always the first to be greeted when a newcomer climbed into a fig tree to join a feeding group of chimpanzees. One day I actually saw him driving another chimp from her nest in order to take it over for himself. It was nearly dark when I observed this from the Peak. The young female had constructed a large leafy nest and was peacefully lying there, curled up for the night. Suddenly Goliath swung up on to the branch beside her and, after a moment, stood upright, seized an overhead branch and began to sway it violently back and forth over her head. With a loud scream she leapt out of bed and vanished into the darkening undergrowth below; Goliath calmed instantly, climbed on to her vacated nest, bent over a new branch, and lay down. Five minutes later the ousted female was still constructing a new bed in the last glimmers of dusk.

William, with his long scarred upper lip and his drooping lower lip, was one of the more subordinate males in his relationships with the other chimpanzees. If another adult male showed signs of aggression towards him, William was quick to approach with gestures of appeasement and submission, reaching out to lay his hand on the other, crouching with soft panting grunts in front of the higher-ranking individuals. During such an encounter he would often pull back the corners of his lips and expose his teeth in a nervous grin. Initially William was timid in camp also. When I offered him a banana in my hand for the first time he stared at it for several moments, gently shook a branch in his frustration, and then sat, uttering soft whimpering sounds, until I relented and put the fruit on the ground.

It took me much longer to determine David Greybeard's position in the dominance hierarchy. In those early days I only knew that he had a very calm and gentle disposition: if William or some youngster approached him with submissive gestures David was always quick to respond with a reassuring gesture, laying his hand on the other's body or head, or briefly grooming him. Often, too, if Goliath showed signs of excitement in camp – if I approached too closely, for instance – David would reach out and lay his hand gently in his companion's groin or make a few brief grooming strokes on Goliath's arm. Such gestures nearly always seemed to calm the more dominant male.

*     *     *

It was during these weeks that Hugo arrived at the Gombe Stream. I had, at last, agreed that a professional photographer be allowed to come and photograph the chimpanzees, and Louis had recommended Hugo. The National Geographic Society had accepted the advice and given Hugo funds for filming me and the chimpanzees – partly to obtain a documentary film record of as much behaviour as possible, and partly in the hope of being able to prepare a lecture film for their members.

Hugo was born in Indonesia, educated in England and Holland and, like myself, had always loved animals. He chose photography as a career on the assumption that he would, somehow and sometime, get to Africa and film wild animals. After two years of working in a film studio in Amsterdam he was accepted by Armand and Michaela Denis to help with the filming of their well-known TV programme *On Safari*. He arrived in Africa, in fact, one year after I did.

Whilst he was working for the Denises, Hugo got to know the Leakeys for they were practically neighbours, and after two years he agreed to make a lecture film for the National Geographic Society on Louis's work at the Olduvai Gorge. It was during this period that Louis decided that Hugo would be just the person to go to the Gombe Stream for he realised that Hugo was not only an excellent photographer, but also that he had a very real love and understanding of animals. Louis wrote to tell me about Hugo and his abilities: at the same time he wrote to Vanne telling her that he had found someone just right as a husband for Jane!

I still felt some apprehension as to how the chimpanzees would tolerate a man with a load of photographic equipment, but I realised the importance of getting a documentary film record of the chimpanzees' behaviour. Also there was David Greybeard – I did not anticipate that David would be too upset by the arrival of a stranger.

On Hugo's first morning at the Gombe Stream, David Greybeard arrived very early in camp, for he had nested nearby. I had decided that it might be best if David first got used to the new tent, and then to its occupant; so, whilst David ate his bananas, Hugo remained inside, peering out through the flaps. David scarcely glanced at the tent until his meal was finished: then he walked deliberately over, pulled back one of the flaps, stared at Hugo, grunted and plodded away in his usual leisurely manner.

To my surprise Goliath, and even timid William, who arrived together a short while afterwards, also accepted Hugo with very little fuss. It was as though they regarded him as merely part of the furniture of the camp. And so, on his first day, Hugo was able to get some excellent film of interactions – greetings, grooming and begging for food – between the three males. And on his second day, Hugo was able to get some even more remarkable film – shots of chimpanzees eating a monkey.

Since that memorable day when I first watched David Greybeard feeding on the carcass of a young bushpig, I had only once seen meat-eating again. On that occasion the prey had been a young bushbuck – but, again, I could not be sure that the chimps had caught it themselves. But on this third occasion Hugo and I actually witnessed the hunt and kill.

It happened most unexpectedly. I had taken Hugo up to show him the Peak, and we were watching four red colobus monkeys that were, it seemed, separated from their troop. Suddenly an adolescent male chimpanzee climbed cautiously up the tree next to the monkeys and moved slowly along a branch. Then he sat down. Three of the monkeys, after a moment, jumped away – quite calmly it seemed. The fourth remained, his head turned towards the chimp. A second later another adolescent male chimp climbed out of the thick vegetation surrounding the tree, rushed along the branch on which the last monkey was sitting, and grabbed it. Instantly several other chimps climbed up into the tree and, screaming and barking in excitement, tore their victim into several pieces. It was all over within a minute from the time of capture.

We were too far away for Hugo to film the hunt – and anyway, it happened so suddenly that he could hardly have expected to do so even had we been close enough. But he did get some film of the chimpanzees eating their share of the kill, though from very far away to be sure.

After such a startlingly auspicious beginning, however, Hugo's luck changed. True, he was able to get a great deal of excellent ciné film and still photographs of David, William and Goliath. But he needed, for his documentary record, more than that. He needed film of as many aspects of the chimpanzees' life as possible – their life in the mountains and forests. And most of them fled from Hugo just as, two years earlier, they had fled from me; even Goliath and William distrusted him when they met him in the forest.

As I had done prior to Judy's visit, I had constructed a few ramshackle hides for Hugo, close to trees that I expected would bear fruit. I had even stuck empty bottles through the walls to try to get the chimps used to the sight of camera lenses. Somehow, though, they instantly detected the difference when they spied real lenses and, if they did arrive in the tree, they usually stared intently towards the hide and vanished silently. Poor Hugo – he lugged most of his camera equipment up and down the steep slopes himself so as not to create the added disturbance of taking an African porter with him. He spent long hours perched on steep rocky hillsides or down on the softer earth of the valley floors that always seemed to harbour biting ants. Frequently no chimps came at all: when they did they often left before he could get even a foot of film.

However, it seemed that the chimps, having very slowly got used to one white-skinned ape in the area, and having then had the chance to study a second, very similar-looking ape in the person of my sister, took a relatively short time to accept a third. And David Greybeard hastened the process. Occasionally, when he saw Hugo or me, he would leave his group and come to see whether, by any chance, we had a banana: the other chimps, of course, watched intently.

When David, William and Goliath had started visiting camp I had soon discovered that they loved chewing on cloth and cardboard – sweaty garments, presumably because of their salty flavour, were the most sought after. One day, when Hugo was crouched in a small hide by a large fruiting tree, a group of chimps climbed up and began to feed. They did not, it appeared, notice him at all. Just as he was beginning to film he felt his camera being pulled away from him. For a moment he couldn't imagine what was going on – then he suddenly saw a black hairy hand pulling at the old shirt which he had wrapped around his camera to try to camouflage the shiny surfaces of the lenses. Of course it was David Greybeard – he had walked along the valley path behind Hugo and then, when he got level with the hide, spied the tempting material. Hugo grabbed hold of one end and engaged in a frenzied tug-of-war until the shirt finally split and David plodded off to join the group in the tree with his spoils in one hand. The other chimps had watched these proceedings with apparent interest, and after that they tolerated Hugo's filming even though, during the struggle with David, most of the hide had collapsed.

In fact, it was only a month after he arrived before most of the chimpanzees accepted not only Hugo, but also his clicking, whirring cameras – provided he kept still and did not move around. But the rains, once again, started early, and just as the weather had consistently ruined Judy's few photographic opportunities the year before, so it was with Hugo. Day after day he sat in his hides, the sun shining, the lighting perfect – but with no chimpanzees to film. Then a group arrived, everything was just right – and, almost as though pre-arranged, it began to rain.

Nevertheless, Hugo did, over the weeks, get some first-rate material on chimpanzee behaviour in the mountains: also, of course, he continued to film interactions between David, Goliath and William in camp.

One of my problems, from the time when I first tried to tempt David and the others to my camp with bananas, had been that of the baboons. There was never a day when the troop did not pass through camp, and some of the males often hung around on their own, as old Shaitani had done, in the hope that bananas would appear. One day, when David, William and Goliath were sitting around a large pile of bananas, a particularly aggressive male baboon suddenly ran at them. William retired from the fray almost at once, his lips wobbling in his agitation, and, with the few bananas he had managed to seize, watched the ensuing battle from a safe distance. David also ran from the baboon when first it threatened him, but he then approached Goliath (who was calmly ignoring the commotion and finishing his bananas) and flung his arms around his friend. Then, as though made brave by the contact, he turned to scream and wave his arms at the baboon. Again the baboon threatened him, lunging forward, and again David rushed to embrace Goliath. This time Goliath responded; he got up, ran a few steps towards the baboon, and then leapt time and time again into the air, in an upright posture, waving his arms and uttering his fierce *waa* or threat bark. David Greybeard joined in, but it was noticeable that he kept a few feet behind Goliath from the start. The baboon retreated, but a moment later, avoiding Goliath, it rushed to lunge and slap at David.

This happened again and again. Goliath leapt at the baboon and the baboon, avoiding him with ease, managed each time to hit out at David, no matter how the latter tried to hide behind his friend. Eventually it was David and Goliath who retreated, leaving the baboon to grab the spoils of victory and rush off to a safer distance. Hugo managed to film the entire incident and, to this day, it remains one of the best records of an aggressive encounter between chimpanzees and baboons.

\*     \*     \*

The National Geographic Society had agreed to finance Hugo's work at the Gombe Stream until the end of November so that he would be able to film the chimpanzees using tools during the termite season. I expected this to start, as in previous years, during October. But, although Hugo and I daily examined different termite mounds, we saw no signs of any activity until early November. Then, when Hugo had only just over two weeks longer at the Gombe, the termites finally began to co-operate. One day, when Hugo made his pilgrimage to his favourite nest near camp, he saw a few moist spots. He scratched away the newly sealed-over passages, poked a grass down and, to his delight, felt the insects grip on. But the chimpanzees, perversely, showed no desire to eat termites. During the following week David, William and Goliath passed the heap frequently but never paused to examine it. Hugo grew desperate. He even led David Greybeard to the termite nest one day, walking along ahead of him with a banana in one hand and, whilst the chimp sat there eating the fruit, offered him a straw-full of clinging termites. David glanced at them and then hit the straw from Hugo's hand with a soft threat bark.

However, during Hugo's last ten days, the chimpanzees finally demonstrated their skill in tool-using and tool-making. Hugo was able to film and take still photographs of David, William and Goliath working at the camp termite heap. It was exciting material, and Hugo hoped that it would help him to persuade the Geographic Society to let him return the following year and continue filming the chimpanzees.

When Hugo left at the end of November I was alone again. I still was not lonely, yet I was not as completely happy in my aloneness as I had been before he came. For I had found in Hugo a companion with whom I could share not only the joys and frustrations of my work, but also my love of the chimpanzees, of the forests and mountains, of life in the wilderness. He had been with me into some of the wild, secret places where, I had thought, no other white person would ever tread. Together we had roasted in the sun and shivered under polythene in the rain. In Hugo I knew I had found a kindred spirit – one whose appreciation and understanding of animals was on a level with my own. Small wonder that I missed him when he was gone.

I shall always remember Christmas at the Gombe Stream that year. I bought an extra large supply of bananas and put them around a little tree which I had decorated with silver paper and cotton wool. Goliath and William arrived together on Christmas morning and gave loud screams of excitement when they saw the huge pile of fruit. They flung their arms around one another and Goliath patted William again and again on his wide-open screaming mouth, whilst William laid one arm over Goliath's back. Finally they calmed down and began their feast, still uttering small squeaks and grunts of pleasure, somewhat muffled and sticky through their mouthfuls of banana.

David arrived much later, on his own. I sat close beside him as he ate his bananas. He seemed extra calm, and after a while I very slowly moved my hand towards his shoulder and made a grooming movement. He brushed me away – but so casually that, after a moment, I ventured to try again. And that time he actually allowed me to groom him for at least a minute. Then he gently pushed my hand away once more. But he had let me touch him, tolerated physical contact with a human – and he was a fully adult male chimpanzee who had lived all of his life in the wild. It was a Christmas gift which I shall always treasure.

I had invited some of the Africans and their children for tea. At first the children were very nervous and ill at ease, but when I produced paper hats and balloons and a few little toys they were so excited that their reserve was soon gone and they were laughing and chasing around and thoroughly enjoying themselves. Iddi Matata, in his stately way, was entranced by the balloons too.

When the party was over I felt a need to go up to the Peak for an hour or so by myself before darkness fell. Then I hurried down to enjoy the Christmas dinner which Dominic had been talking about for days. He and Hugo had planned it, in minute detail, from the stuffing of the chicken to the custard for the plum pudding, before Hugo left. It was dark when I returned, my mouth watering in anticipation – only to find that Dominic, meanwhile, had been celebrating Christmas in his own way. Laid out on my table was one unopened tin of bully beef, an empty plate, a knife and a fork. And that was my Christmas dinner. When I asked Dominic about the chicken and all the rest of it he only laughed uproariously and repeated, again and again, 'To-morrow.' Then he went off, still giggling weakly, to the four-gallon can of local beer which, I later heard, had been brought down by a well-wisher from Bubango, the village on the mountain-top. I couldn't help chuckling as I ate my frugal repast; and Dominic certainly made up for it when he cooked the meal on Boxing Day, despite his king-size hangover!

Shortly after Christmas I had to leave the Gombe Stream myself for another term at Cambridge. My last two weeks were sad, for William fell ill. His nose ran, his eyes watered and he constantly coughed – a dry hacking cough that shook his entire body. The first day of William's illness I followed him when he left camp for, by that time, I was able to move about the forests with both William and David, although Goliath still threatened me if I tried to follow him. William went a few hundred yards along the valley, climbed into a tree, and made himself a large and leafy nest. There he lay until about three in the afternoon, wheezing and coughing and sometimes, apparently, dozing. Several times he actually urinated whilst lying in his bed – behaviour so unusual that I knew he must be feeling very bad indeed. When he finally got up he fed on a few bits and pieces of leaf and vine, wandered slowly back to camp, ate a couple of bananas, and then climbed into a tree just beside my tent and made another nest.

That night I stayed out for there was a moon. At about one o'clock heavy clouds invaded the sky from the East; soon it began to rain. I was crouched some way up the steep slope of the mountain at a slightly higher level than William's nest, and when I shone my powerful torch in his direction I could just make out his huddled figure, sitting up in his wet bed with his knees up to his chin and his arms round his legs. For the rest of the night it rained on and off, and the silences between the pattering of the raindrops were punctuated by William's hacking cough. Once at the start of a really heavy deluge William gave a few rather tremulous pant-hoots and then was silent.

When he climbed down in the morning I saw that, every few moments, his body shook with violent spasms of shivering. As he shivered his long slack lips wobbled, but it was no longer funny. I longed to be able to wrap a warm blanket around him and give him a steaming hot toddy, but all I could offer were a few chilly bananas.

I was with William almost all the time for a week. He spent his days in the vicinity of camp, and many hours lying in different nests. Several times he joined David or Goliath for a while, but when they moved off up the mountain-side he turned back as though he couldn't face a long journey.

One morning I was sitting near William a little way up the mountain above camp when a boat arrived with some visitors from Kigoma for, by then, David Greybeard's fame had spread and people sometimes came for a Sunday picnic in the hope of seeing him. I should, of course, have gone down to say hallo, but I had become so attuned to William that I almost felt myself the chimps' instinctive distrust of strangers. When William moved down towards the tents I followed him: when he sat in the bushes opposite my camp I sat beside him. Together we watched the visitors: they had coffee and chatted for a while and then, since there was no sign of David, they left. I often wondered what they would have thought if they had known that I was sitting there, with William, peering at them as though they had been alien creatures from an unknown world.

One morning, two days before I had to leave, William stole a blanket from Dominic's tent. He had been sitting chewing on it for a while when David Greybeard arrived and, after eating some bananas, joined William at the blanket. For half an hour or so the two sat peacefully side by side, each sucking noisily and contentedly on different corners. But then William, like the clown he so often appeared to be, put part of the blanket right over his head and made groping movements with his hands as he tried to touch David from within the strange darkness he had created. David stared for a moment, and then patted William's hand. Presently the two wandered off into the forest together, leaving me with the echo of a dry hacking cough and the blanket lying on the ground. I never saw William again.

I give David a banana (*Copyright National Geographic Society*)

Figan does a rain dance

Hugo is an all-weather photographer

When David was attacked by this baboon he ran to Goliath for reassurance. During the ensuing battle Goliath repeatedly lunged at the baboon while David sheltered behind him (See p. 176)

*Chapter 7*  Flo's Sex Life

Sex appeal, that strange mystery, that radiation of a certain indefinable IT, is a phenomenon equally inexplicable and just as obvious amongst chimpanzees as humans. Old Flo, bulbous-nosed and ragged-eared, incredibly ugly by human standards, undoubtedly has more than her fair share of it. At one time I thought it was simply because she was an old, and therefore experienced, female that the males got so excited when she became sexually attractive. Now I know better, for there are some old females who are almost ignored at such times, and some young ones who are courted as frenziedly as Flo.

When a female chimpanzee comes into heat – or into oestrus as a scientist would say – the sex skin of her genital area becomes swollen. This pale pink swelling is larger in some females than in others – it may attain the size of a three-pint pudding bowl but is often considerably smaller. A swelling usually persists for about ten days before becoming flabby and wrinkled and then shrinking away to nothing again and, normally, it occurs at a point mid-way between menstrual periods which, in the female chimp, occur about every thirty-five days. It is during her period of swelling that a female – whom I then refer to, frivolously, as a 'pink lady' – is courted and mated by the males.

By the time I got back to the Gombe Stream after my second sojourn at Cambridge – together with Hugo who had, after all, persuaded the National Geographic Society that we needed more film on chimpanzee behaviour – Flo and two of her three offspring had become fairly regular visitors to camp.

Flo's daughter Fifi was about three and a half years old then. She still suckled from her mother for a few minutes every two or three hours, and jumped occasionally on to Flo's back, particularly if she was nervous or startled. I knew, too, that she still shared her mother's nest at night. Figan, Fifi's senior by some four years, had just attained puberty. Some young males of this age are fairly independent of their mothers, but Figan spent most of his days travelling about with Flo and Fifi. Faben, Flo's eldest known son, was seldom seen with his family that year: he was an adolescent of about eleven years at the time.

When Hugo and I first returned to the Gombe Stream, Flo, Fifi and Figan were, of course, still rather apprehensive when they came to camp. They spent most of the time lurking in the thick bushes around the clearing, only emerging to seize the bananas we put out for them. Gradually, however, they relaxed, particularly when David or Goliath was with them, and then they spent longer and longer out in the open. I still spent most of my time climbing the mountains, but when most of the chimpanzees moved far to the north or the south, then I often stayed in camp, hoping that Flo would come.

Even in those days Flo looked very old. She appeared frail, with but little flesh on her bones, and thinning hair that was brown rather than black. When she yawned we saw that her teeth were worn right down to the gums. But we soon found that her character by no means matched her appearance: she was aggressive, tough as nails, and easily the most dominant of all the females at that time.

Flo's personality will become more vivid if I contrast it with that of another old female, Olly, who also began to visit camp at that time: for Olly, with her long face and loose wobbling lips so sadly reminiscent of William, was remarkably different. Flo, for the most part, was relaxed in her relations with the adult males; often I saw her grooming in a close group with two or three males out in the forests, and in camp she showed no hesitation in joining David or Goliath to beg for a share of cardboard or bananas. Olly, on the other hand, was tense and nervous in her relationships with others of her kind. She was particularly apprehensive when in close proximity to adult males, and her hoarse, frenzied pant-grunts rose to near hysteria if high-ranking Goliath approached her. She had a large pendulous swelling in the front of her neck which looked exactly like a goitre. It may, in fact, have been one for they are not uncommon amongst African women in the area; and if so it might account for much of her nervous behaviour.

Olly tended to avoid large groups of chimps and often wandered about with only her two-year-old daughter Gilka for company. Sometimes, too, she was accompanied by her eight-year-old son Evered – indeed, it was he who first led his mother to our camp after coming several times himself with David and Goliath. Often Olly and Flo travelled about together in the forests, and all four children were playmates of long standing.

For the most part the relationship between Flo and Olly was peaceful enough, but if there was a single banana lying on the ground between them, the relative social status of each was made very clear: Flo had only to stick a few of her moth-eaten hairs on end for Olly to retreat, pant-grunting and grinning in submission.

Once, when a game between Flo's son Figan and Olly's son Evered turned into a serious squabble – as happens only too often when young males play together – Flo rushed over at once, her hair on end, in response to Figan's loud screams. She flew at Evered and rolled him over and over until he managed to escape and run off screaming loudly. Olly hurried up too, uttering threatening barks and looking extremely agitated, but she did not dare join in and so contented herself, when all was over, with approaching and, as though to appease the dominant female, laying a hand gently on Flo's back.

Flo was a far more easy-going and tolerant mother than Olly. When Fifi begged for food, whimpering and holding out her hand, Flo nearly always let the child take a banana – sometimes she actually held one out to her. True, there were a few occasions when Fifi tried to take her mother's last banana, and Flo objected; then mother and daughter rolled about on the ground as they fought fiercely for the fruit, screaming and pulling at each other's hair. But such incidents were rare. Gilka would never have dared to try such a thing; indeed, we seldom saw her even begging for Olly's bananas, and when she did, she was usually ignored. She seldom got more than a taste of the skins until she got tame enough to come right up to us so that we could smuggle her a whole banana. And even then, Olly often rushed up and wrenched the fruit away from her daughter.

Despite Flo's normally relaxed relationship with the adult males, she did not usually compete with David or Goliath for bananas; rather she waited until they had taken an armful before she

83

ventured to try and get some for herself. And so, one morning in July 1963, Hugo and I were surprised to see Flo rushing up to join in as Goliath and David approached a pile of bananas. And then we saw that she was pink – in other words, she was flaunting a large sexual swelling.

After seizing a heap of bananas, but before taking a single bite, Goliath stood upright with all his hair on end, stared at Flo and swaggered from foot to foot. As Flo approached, clutching some bananas herself, Goliath raised one arm in the air and made a sweeping downward gesture through the air with his hand. Flo crouched to the ground, presenting Goliath with her pink posterior, and he mated her in the typical nonchalant manner of the chimpanzee, squatting in an upright position, one fruit-laden hand laid lightly on Flo's back and the other resting on the ground beside him.

Chimpanzees have the briefest possible intercourse – normally the male remains mounted for only ten to fifteen seconds. Nevertheless, before Goliath had done with Flo, Fifi was there. Racing up she hurled herself against Goliath, shoving at his head with both her hands, trying to push him off her mother. I expected Goliath to threaten the child, to hit at her – or at least to brush her aside. Instead he merely turned his head away and appeared to try to ignore Fifi altogether. As Flo moved away, Fifi followed, one hand laid over her mother's swelling, looking back over her shoulder at Goliath who sat eating his bananas. For a moment Fifi stayed close to Flo, and then she went off to try and scrounge some fruit for herself.

A few minutes later David Greybeard approached Flo with his hair bristling. Sitting on the ground, he shook a little twig, staring at her as he did so. Flo instantly ran towards him, turned and crouched to the ground: again Fifi raced up and pushed and shoved at David as hard as she could. And David also tolerated her interference.

After this the group settled down. David groomed Flo for a while and then, as it was a hot morning, lay down to snooze. Goliath followed suit, and everything was peaceful. Presently we saw Evered move slightly away from the group, looking over his shoulder at Flo who was watching him. Then he squatted down with his shoulders hunched and his arms held slightly out from his body.

This was the typical adolescent male courtship posture, and Flo responded at once, approaching and presenting as she had for the adult males. Goliath and David both looked up, but otherwise ignored the incident. Fifi, as before, ran up and pushed at Evered, and he too ignored her.

The next day Flo arrived very early in the morning. Her suitors of the day before were with her: once again they courted and mated with her before eating their bananas; once again Fifi rushed up each time and pushed at them. And then, from the corner of his eye, Hugo saw another black shape in the bushes. As we peered we saw another, and another – and another. Quickly we withdrew into the tent and looked through binoculars into the vegetation. Almost immediately I recognised old Mr McGregor. Then I recognised Mike and J.B. Also there were Huxley, Leakey, Hugh, Rodolf, Humphrey – just about all the adult males I knew. And there were some adolescents, females and youngsters in the group as well.

We remained inside the tent and presently Flo moved up into the bushes and there was mated by every male in turn. Always Fifi got there in time and tried to push the suitor away. Once she succeeded: she jumped right on to Flo's back when Mr McGregor was mating her mother and pushed so hard that he lost his balance and tumbled back down the slope.

For the next week Flo was followed everywhere by her large male retinue. It was impossible for her to sit up or lie down without several pairs of eyes instantly swivelling in her direction, and if she got up to move on, then the males were on their feet in no time. Every time there was any sort of excitement in the group – when they arrived at a food source, when they left their nests in the morning, when other chimps joined them – then, one after the other, all the adult males mated Flo. We saw no fighting over this very popular female: each male simply took his turn. Only once, when David Greybeard was mating Flo, did one of the other males show signs of impatience: irascible J.B. started to leap up and down as he swayed a large low branch so that the end beat down on David's head. But David merely pressed himself close to Flo and closed his eyes – and J.B. did not attack him.

The few adolescent males of the group, however, did not stand a chance. Normally these youngsters are able to take their turn provided they wait until a flurry of sexual excitement has died down and the adult males are calm and satiated. Then, if an adolescent hunches his shoulders or shakes a branch at the female from a discreet distance, she will usually move towards him in response. The older males, whilst they may glance at the mating pair, seldom object to these expressions of youthful passion. Flo, however, was an exceptionally popular female. We did sometimes see an adolescent hunching at Flo from behind some tree but, though Flo herself usually got up and wandered towards him, several of the adult males instantly followed. It was as though they feared that Flo might try to escape. The close proximity of his elders effectively nipped the amorous intentions of the adolescent in the bud and he would retreat hastily to gaze at Flo from a safer distance.

Once we saw Evered sitting some way from Flo and, every so often, glancing from her towards one or other of the adult males. It seemed that he was torn between desire and caution. He took a few tentative steps towards Flo and then suddenly went off at a tangent, flicking at pieces of rock, throwing little handfuls of grass in the air, kicking at stones with his feet. Then he sat down and for the next ten minutes occasionally shook a branch as though to relieve his feelings.

On the eighth day of her swelling Flo arrived in camp with a torn and bleeding bottom. The injury must have just occurred: within another couple of hours her swelling had gone. She looked somewhat tattered and exhausted by then and we were relieved, for her sake, that everything was over. At least, we thought it was over, for normally a swelling only lasts about ten days. But five days later, to our utter astonishment, Flo was fully pink again: she arrived, as before, with a large following of attendant males. This time her swelling lasted for three consecutive weeks, during which time the ardour of her suitors did not appear to abate in any way.

It was during this second period of pinkness that we noticed that a strange sort of relationship had grown up between Flo and one of her suitors – a relationship of a type we have never seen since. The male was Rodolf (his real name is Hugo, but two Hugo's in one book become too confusing, and so I have given him Hugo's second name). Rodolf, in those days, was a high-ranking and enormously big and powerful chimpanzee, and he became Flo's faithful escort. He walked everywhere just beside or behind her, he stopped when she stopped, he slept in the nest closest to hers. And it was to Rodolf that Flo often hurried when she was hurt or frightened during those weeks and he would lay his hand reassuringly on her or, sometimes, put one arm around her. Yet he did not protest in any way when other males mated with Flo.

During the final weeks Fifi became increasingly nervous of the males; perhaps, finally, she had been threatened, or even attacked, by one of Flo's suitors. Whatever the reason, her gay assurance in life was gone for a while. She stayed farther and farther away from the centre of a group during any sort of excitement and she completely stopped interfering with her mother's sex life. She didn't even dare join a group that was loading up with bananas in camp – she who, a couple of weeks earlier, had actually taken fruits from the hands of the males whom she had tried to push from Flo.

The fact that Flo's milk had, quite obviously, dried up with the appearance of her sexual swelling, may have affected Fifi's behaviour too, for a quick suck of milk, more than anything else, seems to calm an infant chimpanzee. If one of the males threatened Fifi she could still run to Flo, still be embraced by her mother – but where was the comfort of a sudden flow of warm milk? Whenever the group was resting and peaceful, during those hectic three weeks, Fifi went close to Flo and either groomed her or simply sat beside her with one hand on her mother's body. When the group was on the move Fifi, instead of running jauntily along ahead of Flo, or bouncing along behind, took to riding again as though she were an overgrown baby. Not only did she frequently perch ridiculously on her mother's back, but she sometimes actually clung on in the ventral position under Flo's tummy with her back bumping along the ground.

One day Flo, with Fifi on her back, came into camp alone. Her fabulous swelling had gone, shrivelled into a limp flap of wrinkled skin. She looked worn out, faded, incredibly tattered after her strenuous five weeks. There were two extra pieces torn from her ears and a variety of cuts and scrapes all over her body. That day she just lay around camp for several hours, looking utterly exhausted. She and Fifi left, as they had arrived, on their own.

The following day a group of males was in camp when Flo appeared plodding along the path. The moment they spotted her they leapt up, their hair on end, and charged off to meet her. Flo, with a hoarse scream, rushed up the nearest palm tree. The males ran on with David Greybeard, for once, in the lead. They stopped under her tree and gazed up for a moment before David, slowly and deliberately, began to climb the trunk. What took place we shall never know, for leaves hid the two from sight. A few moments later David reappeared and climbed slowly down the trunk again. He walked past the other five males and plodded back to camp. A moment later Flo reappeared and moved down towards the waiting group. She hesitated for a second before finally stepping to the ground. Then, crouching slightly, she turned and presented her shrivelled posterior to her ex-suitors. Goliath carefully inspected the flabby skin, poking around for a moment and intently sniffing the end of his finger. Then he followed David back to camp and it was Leakey's turn to inspect Flo. After Leakey came Mike, and then Rodolf, and finally old Mr McGregor. Then they all trailed back to their interrupted meal, leaving Flo standing on the path staring after them. Who can tell what she was thinking?

After that Rodolf again attached himself to the old female and, for the next fortnight, continued to travel around with her and her family – Figan, who had stayed away from his mother during most of her pinkness, was back with her also. One day, about a week after the last vestiges of Flo's swelling had subsided, Rodolf, who had been grooming her, suddenly pushed her roughly to her feet and feverishly inspected her bottom, sniffing his finger again and again with an eager glint in his eyes. But, obviously, her hormonal secretions gave no indication of imminent pinkness, and after a few moments he permitted Flo to sit down again and continued to groom her. We saw him behave thus on three different occasions. But Rodolf would have to wait close on five years before Flo again went pink.

When Flo became sexually
attractive and weaned her
daughter, Fifi reverted to infantile
behaviour

Rodolf escorted Flo but did not
stop other males mating with her

Melissa with Goblin

Goliath

Mike

McGregor

Worzle                                    Leakey

Chimpanzees can undoubtedly recognise each other from their voices alone

Eventually the chimpanzees let me follow them into the forest

*Chapter 8*  The Feeding Station

There were two major results of Flo's memorable pinkness. Firstly, she conceived. Secondly, the large number of chimpanzees who had accompanied her during those five weeks had all become accustomed to camp and bananas. They continued to visit us even when Flo was no longer the lure. And so it became worth while to think of setting up a feeding station on more permanent lines: a feeding station which would tempt the nomadic chimpanzees to camp whenever they happened to be in the vicinity so that we should be able to make reasonably regular observations on a number of different, known individuals.

First of all we had to try and think of better ways of offering bananas than merely spraying them about on the ground. For one thing an adult male, if he has the chance, will eat fifty or more bananas at a sitting: for another we were having more and more trouble with the baboon troop. These were problems which took us another six years to solve, but at least, when Hugo and I had to leave at the end of that year, we had made a beginning. With Hassan's help we made a number of concrete boxes with steel lids opening outwards. These boxes were sunk into the ground and the lids held shut by wires attached to handles some distance away. When pins holding the handles were released the wires became slack and the lids fell open.

Some of these boxes were installed by the time Kris Pirozynski, a young Polish mycologist, arrived in early December to study the micro-fungi of the Gombe Stream area. During the four months that Hugo and I would be away, Kris had agreed to look after camp and keep an eye on the chimps. Hassan and Dominic would both stay on to help him, and Dominic was excited at the prospect of making notes on the chimps each day.

By this time Hugo and I were very much in love. But was this, we asked ourselves, simply the outcome of being thrown together in the wilds, far from other European society? Would our feeling change, perhaps, when we found ourselves back in civilisation again? We didn't think so, but as both of us considered marriage a final step, we resolved to test our love. I was going for a third term to Cambridge. Later, Hugo would be joining me and, together, we would show the chimpanzee film to National Geographic Society members in Washington D.C. We would meet again in the world of men rather than the world of apes: and then we would know. As it turned out, however, we knew the very moment we were separated.

I left Hugo about a week before Christmas. On Boxing Day a cable arrived at my family home in Bournemouth: 'WILL YOU MARRY ME LOVE HUGO.' In fact, I should have received the message on Christmas Eve and, although I replied immediately, poor Hugo had had to leave on safari and didn't get my answer for five whole days – not until he got to a place from where he could telephone through to Nairobi.

We decided to get married in London after my Cambridge term and our American lecture were over. Despite the difficulties of planning a wedding when the parties concerned were scattered between Cambridge, Nairobi, Holland and Bournemouth, Hugo and I agreed afterwards that we had never been to a wedding we enjoyed more! A clay model of David Greybeard crowned the wedding cake, and huge coloured portraits of David and Goliath, Flo and Fifi and other chimps looked down on the reception. And everything, from the dresses of myself and my two little bridesmaids to the arum lilies and daffodils, was white and yellow like the spring sunshine that peeped in and out of the fluffy white clouds. It was sad that Louis, after all his machinations and predictions, was unable to be with us. But he sent a speech on tape, and he was represented by his daughter and by his grand-daughter who was one of the bridesmaids.

\*     \*     \*

Three weeks before our wedding we had received letters telling us that Flo had borne a son. We had not, of course, been able to change our wedding plans, but we did cut our honeymoon to only three days in order to get back to the Gombe Stream as quickly as possible.

When we finally made it back to the chimps, after battling our way through flooded rivers, making detours around others and, finally, putting the Land-Rover on to a train, Flo's new infant, whom we subsequently called Flint, was already seven weeks old. But he was still incredibly tiny, still quite hairless on the pink underside of his tummy and chest. If I shut my eyes for a moment I can still recapture, six years later, the thrill of that first moment when Flo came close to us with Flint clinging beneath her. As his mother sat, Flint looked round towards us. His small, pale, wrinkled face was perfect, with brilliant dark eyes, round shell-pink ears and slightly lopsided mouth – all framed by a cap of sleek black hair. He stretched out one arm and flexed the minute pink fingers, then gripped Flo's hair again and turned to nuzzle and rootle with his mouth until he located a nipple. Flo helped him, hitching him a few inches higher and into a better position for suckling. He fed for about three minutes and then seemed to sleep. When Flo moved away she carefully supported him, holding one hand under his back and walking on three limbs.

It was Dominic who had first seen Flo with her new baby. On 28th February she had been in camp, still very pregnant: the following day she had appeared with the tiny infant. She had been accompanied, as usual, by Fifi and Figan: both had sat and stared at the baby, and Fifi had spent a lot of time grooming her mother. Subseqently Figan had seemed less interested in his new sibling, whilst Fifi had become increasingly fascinated.

Dominic and Kris had some other interesting news for us as well. A number of new chimpanzees, including several females, had become regular visitors to camp. Goliath was losing his top-ranking position to Mike. Melissa, one of the young females, seemed to be pregnant. And the chimpanzees were becoming more and more impossible in camp. J.B. had learned to dig boxes and wires out of the ground, so that Hassan had had to sink the boxes in concrete and lay expensive piping between box and handle for the wire. Then J.B. had dug up the pipes, so they had been put into concrete too. Figan and Evered had started trying to prize open the steel lids with strong sticks – occasionally, if the wire was too slack, they succeeded. Even worse, from Kris's point of view, more and more of the chimps had followed David's lead and begun wandering into his tent, taking his clothing and his bedding. Finally he had organised things, stowing any material into tin trunks or stout wooden boxes. Then Goliath had started a craze for chewing canvas. Little groups of chimps had sat around tearing up and chewing chair seats, flaps of tents – even Kris's camp-bed had been destroyed. And finally, he told us, during his last few weeks it had become fashionable to chew wood; the back of one of Hassan's home-made cupboards had gone, and the leg of a wooden chair.

There was also some rather alarming news: some of the bolder chimps had begun to raid the huts of the African fishermen and take *their* clothing. We were worried lest someone, trying to defend his property, might anger or scare one of the big males and get badly hurt in consequence. For the fishermen would not realise the extent to which the chimps had lost their fear of men. We discussed the problem far into one night and eventually decided that, as soon as possible, we would move the feeding area farther up the valley.

The move was surprisingly easy. First of all, with Hassan's help, we installed more of the concrete banana boxes at the new site, and when these were ready we transported our tents and equipment to join them. This was all done after nightfall so as to disturb the chimps as little as possible with the comings and goings of the African porters.

There remained only the task of acquainting the chimps themselves with the new arrangements. One morning I was waiting up at the new camp in the hope that some chimpanzees might pass by – in which case I would offer them bananas. Hugo was down at the beach camp, and we were in communication with one another by means of walkie-talkies. At about eleven o'clock Hugo's voice came through announcing that he had a big group down there and was going to try and lead them the half-mile or so to the new area. For a while there was silence, and then I heard him again, sounding frightfully out of breath. Indeed, it was difficult to make out what he was saying, but eventually I gathered that he wanted me to throw out lots of bananas as far along the path towards beach camp as I could, and as quickly as I could.

I rushed about with armfuls of fruit and had just finished my task when Hugo appeared, running along with a box under one arm and a single banana in his other hand. He hurled the one fruit along the path and collapsed beside me, gasping for breath, as the group of chimps which had been following him suddenly saw the bananas all over the path. With piercing shrieks of excitement they hugged and kissed and patted one another as they fell on the unexpected feast. Gradually their calls became muffled by sticky mouthfuls of banana.

Hugo told me that he had shown a banana to David Greybeard who had been among the six mature males of the group, picked up an empty box of the type which we used for storing bananas, and started to run up the steep skiddy path leading to the new camp. He had not really expected the plan to work, but trusting David, uttering loud barks of delight, had started after him – and the others had immediately followed. Hugo admitted that he had been terrified lest the excited chimps should catch up with him, snatch the box from his arms – and discover that it was empty.

Within a short time the other chimpanzees had discovered the new feeding area. After all, these nomads are used to changes in feeding places – first the figs are ripe in one valley, and then in another. So far as they were concerned the bananas, after an exceptionally long fruiting in one area, had become ripe, in their strange underground boxes, somewhere else.

At the new camp, far from the noise and bustle of the fishermen on the beaches, many of the chimps who had seemed tense and apprehensive before the move soon relaxed. In addition a number of newcomers began to come for bananas. This was good because, at that time, some age groups were very badly represented in our group – we only had two juveniles, for instance, and only a few young adult females. As soon as we noticed a strange face peering from a tree we quickly hid in one of the tents and watched through the mosquito netting windows. Then the stranger would only have to face the novelty of the tents and boxes without the added horror of humans walking around. We used to tip big piles of bananas just outside the tent, from store boxes inside, and hope that the stranger would be able to beg a few from one or other of our chimps, or at least gather up and wedge a few skins.

Often these strangers came many times as far as the trees around our camp, watching the strange goings on of their fellows, before they actually ventured out into the open. Sometimes we were imprisoned within the hot stuffy tent for hours on end. But it was worth it.

One day Goliath appeared, some way up the slope, with an unknown pink female close behind him. Hugo and I quickly put out
a pile of bananas where both chimps could see the fruit and hid in
the tent to watch. When the female suddenly saw our camp she
shot up a tree and stared down. Goliath instantly stopped also
and looked up at her. Then he glanced at the bananas. He moved
a short way down the slope, stopped, and looked back at his female.
She had not moved. Slowly Goliath continued down, and this
time the female climbed silently from the tree and we lost sight of
her in the undergrowth. When Goliath looked round and saw that
she had gone he simply raced back and, a moment later, the female
again climbed into a tree, followed by Goliath who had every hair
on end. He groomed her for a while, but every so often he glanced
towards camp. He could no longer see the bananas, but he knew,
of course, that they were there, and as he had been away for about
ten days his mouth was probably watering.

Presently he climbed down and once more walked towards us,
stopping, every few steps, to stare back at the female. She sat
motionless, but Hugo and I both had the distinct impression that
she wanted to escape from Goliath's company. When Goliath had
come a bit farther down the slope the vegetation obviously hid the
female from his view – he looked back and then quickly climbed a
tree. She was still sitting there. He climbed down, walked another
few yards, and then hastened up another tree. Still there. This
went on for a further five minutes as Goliath proceeded towards
the bananas.

When he reached the camp clearing Goliath faced an added
problem – there were no trees to climb and he couldn't see the
female from the ground. Three times he stepped into the open –
then turned and rushed back up the last tree. The female did not
move. Suddenly Goliath seemed to make up his mind and, at a fast
canter, ran over to the bananas. Seizing only one he turned back
and raced to climb his tree again. Still the female sat on the same
branch. Goliath finished his banana and, as though slightly reassured, hastened back to the pile of fruit, gathered up a whole
armful, and hurried back to the tree. But this time the female had
gone: whilst Goliath had gathered the bananas she had climbed
down from her branch, repeatedly glancing towards him over her
shoulder, and vanished silently.

Goliath's consternation was amusing to watch. Dropping all his bananas he raced up to the tree where he had left her, peered all round, and then he too vanished into the undergrowth. For the next twenty minutes he searched for that female; every few minutes we saw him climbing up yet another tree, staring in all directions. But he never found her and finally he gave up, returned to camp and sat, looking quite exhausted, slowly eating bananas. Even so, he kept turning his head to gaze back up the slope.

I remember too when an old mother, whom I knew well in the forests, appeared for the first time at the outskirts of our clearing. She remained in the trees, watching, but her four-year-old son moved on into camp with the rest of the group. To our astonishment he actually came on to the veranda of the tent and, as we crouched inside, hardly daring to breathe lest we scare him, the corner of the flap was lifted and his small face stared in at us. Then, quite calmly, he lowered the flap and resumed his search for banana peels. He was the boldest youngster we ever encountered.

It was during these months that we first realised what an exceptionally gifted chimp was young Figan. As more and more chimps discovered the feeding area we soon found that we did not have sufficient concrete boxes, and it took a long time to get the lids made in Kigoma. It became more and more difficult to ensure that the females and youngsters had their share. And so we took to hiding some fruits, one here, one there, up in the trees; youngsters such as Figan quickly learnt to search for these whilst the adult males were busy loading up at the boxes. One day, some time after the group had been fed, Figan suddenly spotted a banana that had been overlooked – but Goliath was resting directly underneath it. After no more than a quick glance from the fruit to Goliath, Figan moved away and sat on the other side of the tent so that he could no longer see the fruit. Fifteen minutes later, when Goliath got up and left, Figan, without a moment's hesitation, went over and collected the banana. Quite obviously he had sized up the whole situation: if he had climbed for the fruit earlier Goliath, almost certainly, would have snatched it away. If he had remained close to the banana he would probably have looked at it from time to time: chimps are very quick to notice and interpret the eye movements of their fellows, and Goliath would possibly, therefore, have seen the fruit himself. And so Figan had not only refrained from instantly gratifying his desire, but had also gone away so that he could not give the game away by looking at the banana. Hugo and I were, properly, impressed. But there was more to come.

Quite often, when chimpanzees have been resting, if one gets up and walks away without hesitation, the others will follow. It does not need to be a high-ranking individual – a female or a youngster may start such a move. One day, when Figan was part of a large group and, in consequence, had not managed to get more than a couple of bananas for himself, he suddenly got up and walked away. The others trailed after him. Ten minutes later he returned, quite by himself – and, of course, got his share of bananas. We thought this was coincidence – indeed, it may have been on that first occasion. But after this the same thing happened over and over again – Figan led a group away and returned, later, for his bananas. Quite obviously he was doing it deliberately. One morning, after such a manœuvre, he returned with his characteristic jaunty walk, only to find that a high-ranking male had, in the meantime, arrived in camp and was sitting eating bananas. Figan stared at him for a few moments and then flew into a tantrum, screaming and hitting at the ground. Finally he rushed off after the group he had led away earlier, his screams gradually receding in the distance.

That camp was a perfect spot for a newly-married couple. Our tents were pitched in the dense shade of a little palm grove and we looked out over a natural grassy clearing that was brilliant, for over four months, with the scarlet blossoms of ten or more candelabra trees. Gleaming metallic sunbirds of many species flitted about during the day feeding on the nectar, and often, in the evening, a bushbuck picked his way daintily across the grass. At the far end of the clearing the stream ran and there, in the cold mountain water, we washed in the evening. We cooked our own breakfasts and ate a couple of slices of bread for lunch – for Hugo was determined I should not become a skeleton now that I was his wife. And then, as darkness fell, Dominic and Sadiki, the local African we had employed to help with the work of the camp, came up with our supper and tidied camp.

We shall never forget those days. Over and above the growing love which enriched our lives, and the beauty of the mountains and the forests around us, we were sharing work and experience which we both enjoyed more than anything else – watching and learning about animals.

That year, amongst other things, we discovered a new chimpanzee tool. It was when we were out in the forest, watching the old mother Olly with her two offspring, Gilka and Evered. Evered, as he climbed through a tree, suddenly stopped and, with his face close to the bark, peered into what looked like a small hollow. He picked a handful of leaves, chewed them for a moment, took them out of his mouth and pushed them down into the hollow. As he withdrew them we saw the gleam of water. Quickly Evered sucked the liquid from his home-made sponge and poked it down into the hollow once more. At that moment Gilka came up and watched him intently: when he moved away she made a tiny sponge and pushed it into the hollow but it seemed that all the water had gone; she dropped her sponge and wandered off. Later we saw the same behaviour many times, for we made an artificial water bowl in a fallen tree-trunk in camp. Always the chimpanzees first crumpled and chewed the leaves – and that, of course, made the sponge much more absorbent. It is, in fact, another example of tool-*making*.

But of course the most exciting thing, that year, was that we were able to record, on paper and on film, the week-by-week development of a wild chimpanzee infant – Flint. Flo and her family had been well known to us before; now they became an integral part of our lives. We learnt a great deal about their behaviour by objective recording of facts, but we also became increasingly aware of them as individual beings: intuitively we 'knew' things about them which, as yet, we could not begin to define in scientific terms. We began, though indeed 'through a glass darkly', to understand what a chimpanzee really is.

Our only disappointment that year was the fact that we missed the first few weeks of Flint's life, but the birth of a baby to Melissa almost made up for that. The heat of the day was over, and the sun low in the sky, when we first saw the tiny infant. As Melissa came down the slope towards our camp she moved on three limbs, supporting the newborn with one hand. Every so often she stopped and seemed to disentangle something from the undergrowth; when she got closer we saw that this was the placenta, still attached to the baby by the umbilical cord.

Melissa came right up to us, quite unafraid for her infant. She seemed dazed, her eyes not quite focused, her movements slow and uncertain. When one of the mature males arrived Melissa, usually so quick to greet another chimp, so anxious to ingratiate herself with her superiors, ignored him completely. Nor did she follow when, after a short while, he left. She continued to sit, the baby cuddled between her thighs, her feet crossed under his tiny rump, her arm behind his head. For some while we could not see the infant at all and then, as she finished her few bananas, she removed her encircling arm.

The baby's head fell back on to her knees, and Melissa, looking down, stared and stared at the tiny face. Never had we imagined such a funny twisted-up little face. It was comical in its ugliness, with large ears, small rather pursed lips, and the skin incredibly wrinkled and bluish black rather than pink. His eyes were screwed tight shut against the fading light of the sun, and he looked like some wizened gnome or hobgoblin. We christened him Goblin on the spot. Melissa gazed down at her son for fully two minutes before she placed one hand behind his back and set off to make her nest for the night.

Hugo and I followed, keeping well behind. Every fifteen steps or so Melissa stopped and sat for a few moments before moving on, still supporting the infant with one hand, the placenta still trailing. It was dusk when she reached a tall leafy tree and climbed up, and we could hardly see by the time she had finished making her bed. She made a very large one, using her feet and one hand, and it took her eight minutes instead of the usual three to five. Finally she lay down and all was quiet.

We left her then, climbing back down the mountain to our forest home, rather silent as we thought of the young female, bewildered by the miracle of birth as so many other mothers have been throughout the centuries, animal and human alike. For the first time since leaving her own mother, Melissa was sharing her nest with another chimpanzee.

# Chapter 9   Flo and Her Family

Old Flo lay on her back in the early morning sunshine, her belly full of palm nuts, and suspended Flint above her, grasping one of his minute wrists with her large horny foot. As he dangled, gently waving his free arm and kicking with his legs, she reached up and tickled him – in his groin and his neck – until he opened his mouth in the play-face, or chimpanzee smile. Close by Fifi sat, staring at Flint, occasionally reaching out to touch her ten-week-old brother gently with one hand.

Faben and Figan, Flo's two elder sons, played with each other nearby: since Flint's birth two and a half months earlier Faben had begun to move around with his family more and more frequently. Every so often as their game became extra vigorous, I could hear the panting chuckles of chimpanzee laughter.

Suddenly Faben, three or four years Figan's senior, began to play rather roughly, sitting down and kicking with the soles of his feet on Figan's bent head. After a few moments Figan had had enough: he left Faben and, with his jaunty walk, approached Fifi and tried to play with her. But at that moment Flo, gathering Flint to her breast, got up to move into the shade, and Fifi pulled away from Figan to follow her mother. For, since Hugo and I had returned to the Gombe Stream, when Flint was seven weeks old, Fifi had become increasingly fascinated by her new brother.

Flo sat down and began to tickle Flint's neck with small nibbling movements of her worn old teeth, and Fifi, once again, sat close by and reached out to make a few grooming movements on Flint's back. Flo ignored this. Earlier, though, when Flint was under two months, Flo had usually pushed Fifi's hand away each time she tried to touch Flint, and often the only way in which the child had been able to contact the infant had been by solicitously grooming Flo, working ever closer and closer to those places where

Flint's hands gripped his mother's hair. Intently Fifi had groomed around the hands, occasionally briefly fondling the minute fingers and then, with a glance at Flo, hastily returning to her grooming.

Now, however, Flint was older and, for the most part, Fifi was permitted to touch him. As I watched, Fifi began to play with Flint, taking one hand and nibbling the fingers. Flint gave a soft whimper – possibly Fifi had hurt him – and instantly Flo pushed her daughter's hand away and cuddled her infant closely. Frustrated, Fifi rocked slightly to and fro, twisting her arms behind her head and staring at Flint, her lips slightly pouted. It was not long before she reached out, gently this time, to touch him again.

I have always thought that human children become increasingly fascinating as they grow out of the helpless baby stage and begin to respond to people and things. Certainly a chimpanzee baby becomes more attractive as it grows older, not only to its mother and siblings, but to the other members of the community – and to mere human observers. For Hugo and I, the privilege of being able to watch Flint's progress that year remains one of the most delightful of our experiences – comparable only with the joy we were to know much later as we watched our own son growing up.

When Flint was three months old he was able to pull himself about on Flo's body, taking handfuls of her hair, pulling with his arms and pushing with his feet. And, at this time, he began to respond, when Fifi approached, by reaching out towards her. Fifi became more and more preoccupied with him. She began to make repeated attempts to pull him away from his mother. At first Flo firmly prevented this, but even when Fifi persisted, pulling at her brother again and again, Flo never punished her. Sometimes she pushed the child's hand away, sometimes she simply walked away, leaving Fifi rocking slightly, her limbs contorted. And sometimes, when Fifi was extra troublesome, Flo, instead of repulsing her advances, either groomed her or played with her quite vigorously. These activities usually served to distract Fifi's attention, at least temporarily, from her infant brother.

As the year wore on it seemed that Flo, perhaps as a result of playing so often with Flint and Fifi, her two younger children, became more and more playful. Often, as the weeks passed, we saw her playing with both Figan and twelve-year-old Faben, tickling them or chasing with them round and round a tree-trunk, with

Flint hanging on for dear life. On one occasion, in the middle of a romp with Faben, this old female lowered her balding head to the ground, raised her bony rump in the air, and actually turned a somersault. And then, almost as though she felt slightly ridiculous, she moved away, sat down, and began to groom Flint very intently.

When Flint was thirteen weeks old we saw Fifi succeed in pulling him away from his mother. Flo was grooming Figan when Fifi, with infinite caution and many quick glances towards her mother's face, began to pull at Flint's foot. Inch by inch she drew the infant towards her – and all at once he was in her arms. Fifi lay on her back and cuddled Flint to her tummy with her arms and legs. She lay very still.

To our surprise Flo, for the first few moments, appeared to take no notice at all. But when Flint, who had possibly never before lost contact with his mother's body, reached round and held his arms towards her, pouting his lips and uttering a soft 'hoo' of distress, Flo instantly gathered him to her breast and bent to kiss his head with her lips. Flint eagerly sought the reassurance of his mother's breast, suckling for a few moments before turning to look at Fifi again. And Fifi, her hands clasped behind her head, her elbows in the air, stared and stared at Flint.

Ten minutes later Fifi was again permitted to hold Flint for a short while but, once more, the moment Flint gave his tiny distressed whimper Flo rescued him. And Flint, as before, suckled briefly when he regained the security of his mother's arms.

After this not a day passed but that Fifi pulled her infant brother away from Flo. As time passed Flint became accustomed to the arms of his sibling and so she was able to hold him for longer and longer before he uttered the tiny sound that, for the next nine months, would bring Flo hastening to his rescue. Flo even permitted Fifi to carry Flint when the family wandered through the forests.

On those occasions when Flo and her family were part of some big group, however, Flo was more possessive of her infant. Then, if Fifi moved away with Flint, Flo followed, uttering soft whimpers herself, until she caught up with the 'kidnapper' and retrieved her infant. Even then, however, Fifi was not punished; Flo simply reached forward, grabbed hold of her daughter's ankle, and then gathered Flint into her arms. Sometimes Fifi led her old mother a merry dance, round and round trees, under low vegetation where

Flo had to creep almost on her belly – even up into the trees. And sometimes too, as though to prevent Flo from catching hold of her, she walked backwards in front of her mother, grunting softly and bobbing up and down, as though in submission, but not, until she was forced to, relinquishing Flint.

When Flint was very small his two elder brothers, although they sometimes stared at him, paid him little attention. Occasionally Faben, whilst he was grooming with his mother, very gently patted the infant, but Figan, though he was such an integral part of the family, seemed afraid to touch Flint in the early days. If, when Figan and Flo were grooming, the infant accidentally touched Figan as, baby fashion, he waved his arms about, Figan, after a quick glance at Flo's face, seemed to avoid looking at Flint. For Figan, though he was a vigorous adolescent male, still showed great respect for his old mother.

One occasion is vivid in my memory. Fifi had taken Flint and was sitting grooming the infant some ten yards from Flo. Presently Figan approached and sat beside his sister. Flint turned towards him and, with his wide-eyed gaze fixed on Figan's face, reached out to grasp his brother's chest hair. Figan started and, after a quick glance in Flo's direction, raised his hands up and away from the baby. Then he stayed motionless, staring down at Flint, his lips tense. The infant pulled closer and nuzzled at Figan's breast – then all at once seemed afraid of the unfamiliar. Usually, of course, his only contacts were Flo and Fifi, and if he reached towards either of them they always responded by holding him close. With a slight pout Flint turned back to Fifi, but then, as though confused, he again reached to Figan with a soft whimper. At this Flo came hurrying to his rescue and, as she approached, Figan, too, gave low worried cries and raised his hands even higher as in the age-old gesture of surrender. Flo gathered up her infant and Figan lowered his hands slowly as though dazed.

One day, when Flint was just under five months old, Flo got up to go and, instead of pressing Flint to her belly, took his arm in one hand and hoisted him over her shoulder on to her back. There he remained for a few yards before he slipped down and clung to her arm. For a short distance Flo continued, with Flint gripping around her elbow, and then she pushed him back under her tummy. But the next day, when Flo arrived in camp, Flint was clinging precariously to her back, hanging on to her sparse hair with his hands and feet. When Flo left she again pushed her son up on to her back, and again he clung there for a while before sliding down and dangling, from one hand, by her side. This time, after walking thirty yards or so, Flo pushed him once more on to her back. After this Flint nearly always rode on Flo's back or else dangled beside her whilst she walked the mountains; this was not surprising, for all infants, after a certain age, start riding their mothers rather than clinging on beneath. But we were astonished to see that Fifi, when next we saw her take Flint, also tried to push him on to her back. This was surely an example of learning by direct observation of her mother's behaviour.

By the time Flint was five months old he had become an accomplished rider, and only occasionally slipped down to dangle beside Flo as she walked. But if there was any sign of excitement amongst the group, or if Flo was about to move into thick undergrowth, then she always reached back and pushed Flint round so that he clung, as before, underneath. After a while he learned to wriggle round under Flo of his own accord in response to the slightest touch.

It was about the same time as Flint began to ride on Flo's back that we first saw him take a step by himself. For some weeks previously he had been able to stand on the ground balanced on three limbs and clinging to Flo's hair with one hand; and occasionally he had taken a couple of steps in this manner. On this particular morning he suddenly let go of Flo and stood by himself, all four limbs on the ground. Then, very deliberately, he lifted one hand off the ground, moved it forward safely, and paused. He lifted a foot off the ground, lurched sideways, staggered and fell on his nose with a whimper. Instantly Flo reached out and scooped him into her arms. But that was the beginning. Each day after that Flint walked one or two steps farther, but for months he was incredibly wobbly. Constantly he got his hands and feet muddled up and fell – and always Flo was quick to gather him up. Often, indeed, she kept one hand under his tummy as he tottered along.

Just after he began to walk Flint began to try to climb. One day we saw him standing upright, holding on to a tiny sapling with both hands, and gripping it first with one foot and then the other. But he never managed to get both feet off the ground at once, and after a few moments he fell backwards on to the ground. Subsequently he repeated this performance several times, and Flo, as she groomed Fifi, idly held one hand behind his back, preventing further tumbles. A week after his first attempt Flint was able to climb a short way quite easily. Like a human child he found it much harder to get down by himself, but Flo, of course, was very watchful – as indeed was Fifi – and one or other of his guardians reached to rescue him the moment he gave his soft whimper. Flo, in fact, often retrieved him when she noticed that the end of the branch on which he was swinging was beginning to bend, and when Flint himself was perfectly contented. She was equally quick to seize him if she saw any sign of social excitement or aggression amongst other members of the group.

Gradually Flint learnt to control his limbs slightly better when walking, although he still often relied on speed, rather than co-ordination, to get him from one place to another. He began to venture several yards away from Flo – and since any movement away from his mother was intensely exciting, and any excitement set his hair on end, he tottered around like a fluffy black ball, his wide-eyed gaze fixed intently on some object or individual in front of him.

It was at this time that Fifi's fascination for her small brother became almost an obsession. She spent nearly all her day playing with him, grooming him as he slept, or carrying him about with her. Flo, it seemed, was often far from displeased to shed, from time to time, part of her load of maternal responsibility. Provided that Fifi did not carry Flint out of sight, and provided there were no potentially aggressive males nearby, she no longer objected when Fifi 'kidnapped' Flint. Nor did Flo seem to mind if other youngsters approached Flint to play gently with him. But Fifi did! If she suddenly noticed Gilka, or another of her erstwhile playmates, close to Flint, Fifi instantly abandoned whatever she was doing, rushed over, and chased the youngster away, her hair bristling, her arms flailing, her feet stamping the ground. Even chimps much older than herself, provided they were subordinate to Flo, were threatened or even attacked by aggressive Fifi. Presumably she acted on the assumption that if anything went wrong old Flo would hurry to her assistance – and it seemed that the victims of her fury were fully aware of this fact also.

Fifi, however, could not chase Figan or Faban away from Flint and, as the infant grew up, both his elder brothers showed more and more interest in him. Often they would approach and play with him, tickling him or pushing him gently to and fro as he dangled, legs kicking, from a low branch. Sometimes when Figan was playing with Flint, we saw Fifi hurry up and try to initiate a game with Figan; often she was successful. And then, when the game was over, Fifi hurried back to play with Flint herself. Was she, perhaps, practising the same technique of distraction that Flo had used so often on her?

When Flint tottered up to one of the adult males Fifi, of course, could scarcely interfere; she merely sat and stared as David, or Goliath, or Mike reached out and, time and again, patted Flint or gently embraced him. And, as the weeks went by, Flint, like a spoilt human child, wanted more and more attention. One day as he wobbled up to Mr McGregor, the old male got up and moved away. It was not, I think, deliberate – it just happened that he was about to leave. Flint stopped dead, staring with widening eyes at the male's retreating rear, and then, stumbling along with frantic haste, falling again and again on his face, Flint followed. All the time he uttered his soft whimper. Within minutes Flo was rushing to retrieve him. But that was only the start of it, and for the next few weeks Flint was always whimpering along after one or other of the adult males who had not deigned to stop and greet him or who had walked away from the infant for any reason whatsoever. Quite often the male concerned, uneasy perhaps at the little calls following in his wake, stopped or turned back to pat Flint.

When Flint was eight months old he sometimes spent fifteen minutes or so out of contact with Flo as he played or explored, but he was never very far from her. He was a little steadier on his feet, and he was able to join Fifi in some of her slightly rougher games, chasing round and round a grass tuft, or pulling himself on top of her as she lay on the ground and tickling her with his hands and mouth. It was at this time that the termiting season began.

One day, when Flo was fishing for termites, it became obvious that Figan and Fifi, who had been eating termites at the same heap, were getting restless and wanted to go. But old Flo, who had already fished for two hours, and who was herself only getting about two termites every five minutes, showed no signs of stopping. Being an old female, it was possible that she might continue for another hour at least. Several times Figan had set off resolutely along the track leading to the stream, but on each occasion, after repeatedly looking back at Flo, he had given up and returned to wait for his mother.

Flint, too young to mind where he was, pottered about on the heap, occasionally dabbing at a termite. Suddenly Figan got up again and this time approached Flint. Adopting the posture of a mother who signals her infant to climb on to her back, Figan bent one leg and reached back his hand to Flint, uttering a soft, pleading whimper. Flint tottered up to him at once, and Figan, still whimpering, put his hand under Flint and gently pushed him on to his back. Once Flint was safely aboard, Figan, with another quick glance at Flo, set off rapidly along the track. A moment later Flo discarded her tool and followed.

Hugo and I were amazed at this further example of Figan's ingenuity in getting his own way. But had his behaviour really been deliberate? We couldn't, of course, be sure. And then, a few days later, Fifi did exactly the same thing. Subsequently we watched Faben take Flint to his breast after he, too, had tried several times to persuade his mother to follow him away from a termite heap. We had never seen Faben carrying Flint before.

As the termite season wore on there could be no doubt that Flo's elder offspring were kidnapping Flint with the deliberate intent of getting their mother to stop, at least for a while, her endless termiting. We saw all three of them taking Flint in this way on any number of occasions. Of course, they were not always successful. Often Flint dropped off and ran back to his mother of his own accord. And sometimes, if Flo's hole was still yielding a good supply of termites, she hurried to retrieve Flint and then returned to the heap followed by the unsuccessful kidnapper – who usually tried again later.

Flint, of course, was too young to show any interest in eating termites – he now and then sampled a mouthful of fig or banana, but still received virtually all his nourishment from his mother's milk and would continue to do so for another year. He did occasionally poke at a crawling termite with his finger, and he played with discarded grass tools when he was wandering about on a termite heap. Also he began to 'mop' everything. When termites are spilled on to the surface of the heap, older chimps mop them up with the backs of their wrists: the termites become entangled in the hairs and are picked off with the lips. It was soon after the termite season began that Flint started to mop things – the ground, his own legs, his mother's back as he rode along – anything but termites! In fact, though he sometimes gazed intently for a few moments as his mother or one of his siblings worked, he was not really interested in this activity which so absorbed his elders.

Fifi, on the other hand, was a keen termite-fisher, and when Flint, wanting to play with his sister, jumped on to her and scattered the insects from her grass stem, she was quite obviously irritated. She pushed him away roughly time and time again. Of course, Fifi still played with Flint frequently herself when she was not ter-miting, but, almost as though some spell had been broken, she never again showed quite the same fanatical preoccupation with him; she no longer protected him so consistently from social contact with other young chimps.

And so Flint began to enlarge his circle of friends for Fifi, parti-cularly when she was working at a termite heap, often permitted Gilka or one of the other juveniles to approach and play with Flint and she no longer rushed up aggressively every time one or other of the adolescent females carried Flint around or groomed him or played with him. Flint, in fact, was growing up. Even when Fifi did devote all her attention to her little brother she could no longer treat him as her doll, for he had developed a mind of his own. If Fifi wanted to carry him in one direction and he wanted to go some-where else, then he struggled away from her and went his way. Also, of course, he was getting heavier. One day, when Flint was sleeping in her lap and gripping tightly to her hair, it was obvious that he was hurting his sister. Fifi carefully detached first one hand and then the other, but as soon as they were loosened Flint, dis-turbed, gripped on again tightly. Finally, for the first time on record, Fifi carried the infant back to Flo and pushed him in her mother's direction.

When Flint was one year old he was still wobbly on his legs, but he was quick to bounce towards any game that was in progress, and eager to hurry over to greet any newcomer that joined his group. He was, in fact, beginning to take part in the social life of his community: a community which, at that time, was still un-settled as a result of the dramatic rise to overall dominance of Mike. Flint could scarcely have been aware of the battle of wills that had finally led to Goliath's defeat for it had started at the time of his birth; Flint grew up in a world where Mike, undisputedly, was supreme.

*Chapter 10* The Hierarchy

Mike's rise to the number one or top-ranking position in the chimpanzee community was both interesting and spectacular. In 1963 Mike had ranked almost bottom in the adult male dominance hierarchy. He had been the last to gain access to bananas, and had been threatened and actually attacked by almost every other adult male. Indeed, at one time he had appeared almost bald from losing so many handfuls of hair during aggressive incidents with his fellow-apes.

When Hugo and I had left the Gombe Stream at the end of that year, prior to getting married, Mike's position had not changed; yet when we returned, four months later, we found a very different Mike. Kris and Dominic told us the beginning of his story – how he had started to use empty four-gallon paraffin cans more and more often during his charging displays. We did not have to wait many days before we witnessed Mike's techniques for ourselves.

There was one incident that I remember particularly vividly. A group of five adult males, including top-ranking Goliath, David Greybeard and the huge Rodolf, were grooming each other – the session had been going on for some twenty minutes. Mike was sitting on his own about thirty yards from them, frequently staring towards the group, occasionally idly grooming himself.

All at once Mike calmly walked over to our tent and took hold of an empty paraffin can by the handle. Then he picked up a second can and, walking upright, returned to the place where he had been sitting before. Armed with his two cans, Mike continued to stare towards the other males and, after a few minutes, he began to rock from side to side. At first the movement was almost imperceptible, but Hugo and I were watching him closely. Gradually he rocked more vigorously, his hair slowly began to stand erect, and then, softly at first, he started a series of pant-hoots. As he called, Mike got to his feet and suddenly he was off, charging towards the group of males, hitting the two cans ahead of him. The cans, together with Mike's crescendo of hooting, made the most appalling racket: no wonder the erstwhile peaceful males rushed out of the way. Mike and his cans vanished down a track and, after a few moments, there was silence. Some of the males reassembled and resumed their interrupted grooming session, but the others stood around somewhat apprehensively.

After a short interval that low-pitched hooting began again, followed, almost immediately, by the appearance of the two rackety cans with Mike close behind them. Straight for the other males he charged, and once more they fled. This time, even before the group could reassemble, Mike set off again: but he made straight for Goliath – and even he hastened out of Mike's way like all the others. Then Mike stopped and sat, all his hair on end and breathing hard. His eyes glared ahead and his lower lip was hanging slightly down so that the pink inside showed brightly and gave him a wild appearance.

Rodolf was the first of the males to approach Mike, uttering soft pant-grunts of submission, crouching low and pressing his lips to Mike's thigh. Then he began to groom Mike, and two other males approached, pant-grunting, and began to groom him also. Finally David Greybeard went over to Mike, laid one hand on his groin, and joined in the grooming. Only Goliath kept away, sitting on his own and staring towards Mike. It was obvious that Mike constituted a serious threat to Goliath's hitherto unchallenged supremacy.

Mike's deliberate use of man-made objects was probably an indication of superior intelligence. Many of the adult males had, at some time or another, dragged a paraffin can to enhance their charging displays, in place of the more normal branches or rocks; but only Mike apparently had been able to profit from the chance experience and learned to seek out the cans deliberately to his own advantage. The cans, of course, made a great deal more noise than a branch when dragged along the ground at speed, and, after a while, Mike was actually able to keep three cans ahead of him at once for about sixty yards as he ran flat out across the camp clearing. No wonder that males, previously his superiors, rushed out of Mike's way.

Charging displays usually occur when a chimpanzee becomes emotionally excited; when he arrives at a food source, joins up with another group or when he is frustrated. But it seemed that Mike actually *planned* his charging displays – almost, one might say, in cold blood. Often, when he got up to fetch his cans, he showed no visible signs of frustration or excitement – that came afterwards when, armed with his display props, he began to rock from side to side, raise his hair, and hoot.

Eventually Mike's use of paraffin cans became dangerous for he learned to hurl them ahead of him at the close of a charge – once he got me on the back of my head, and once he hit Hugo's precious film camera. We decided to remove all the cans and, for a while, went through a nightmare period since Mike tried to drag about all manner of other objects. Once he got hold of Hugo's tripod – luckily when the camera was not mounted – and once he managed to grab and pull down the large cupboard in which we kept a good deal of food and all our crockery and cutlery. The noise and the trail of destruction were unbelievable. Finally, however, we managed to dig things into the ground or hide them away, and Mike had to resort to branches and rocks like his companions.

By that time, however, his top-ranking status was assured, although it was fully another year before Mike himself seemed to feel quite secure in his position. He continued to display very frequently and vigorously, and the lower-ranking chimps had increasing reason to fear him for often he would attack a female or youngster viciously at the slightest provocation. In particular, as might be expected, a tense relationship prevailed between Mike and the ex-dominant male, Goliath.

Goliath did not relinquish his position without a struggle. His displays also increased in frequency and vigour and he too became more aggressive. Indeed, there was a time, towards the start of this battle for dominance, when Hugo and I feared for Goliath's sanity. After attacking a couple of youngsters and charging back and forth dragging huge branches, he would sit, his hair on end, his sides heaving from exertion, a froth of saliva glistening at his half-open mouth, and a glint in his eyes that, to us, looked not far from madness. We actually had a weld-mesh iron cage built in Kigoma and, when this had been set up in camp, we retreated inside when Goliath's temper was at its worst.

One day, when Mike was sitting in camp, a series of distinctive rather melodious pant-hoots, with characteristic quavers at the close, announced the return of Goliath who, for two weeks, had been somewhere down in the south of the Reserve. Mike responded immediately, hooting himself and charging across the clearing. Then he climbed a tree and sat staring over the valley, every hair on end.

A few minutes later Goliath appeared and, as he reached the outskirts of the camp clearing, he commenced one of his spectacular displays. He must have seen Mike for he headed straight for him, dragging a huge branch. Then he leapt up into a tree near that of Mike and was still. For a moment Mike stared towards him and then he too began to display, swaying the branches of his tree, swinging to the ground, hurling a few rocks and, finally, climbing up into Goliath's tree and swaying the branches there. When he stopped Goliath immediately reciprocated, swinging about in the tree and rocking the branches. Presently, as one of his wild leaps took him quite close to Mike, Mike too displayed, and for a few unbelievable moments both of the splendid male chimpanzees were swaying branches within a few feet of each other until I thought the whole tree must crash to the ground. But an instant later both chimps were on the ground, displaying in the undergrowth. Finally they stopped and sat, staring at each other. It was Goliath who moved next, standing upright as he rocked a sapling; when he paused Mike charged past him, hurling a rock and drumming, with his feet, on the trunk of a tree.

This went on for nearly half an hour: first one male and then the other displayed, and each performance seemed to be more vigorous, more spectacular, than that preceding it. Yet during all this time, apart from occasionally hitting one another with the ends of the branches they swayed, neither chimpanzee actually attacked the other. Suddenly, after an extra long pause, it seemed that Goliath's nerve broke. He rushed up to Mike, crouched beside him with loud, nervous pant-grunts, and began to groom him with feverish intensity. For a few moments Mike ignored Goliath completely: then he turned and, with a vigour almost matching that of Goliath, began to groom his vanquished rival. And there they sat, grooming each other without pause, for over an hour.

That was the last real duel between the two males. From then on

Fifi (*Copyright National Geographic Society*)

Fifi continually wanted to touch her young brother

Ultimately Flo allowed her to carry Flint

Flint reached from Fifi to Figan. When Flo approached, Figan raised his arms as if to show he was not harming the infant (See p. 102)

Faben, Fifi and Figan play around Flo and Flint

Flo resting on the ground

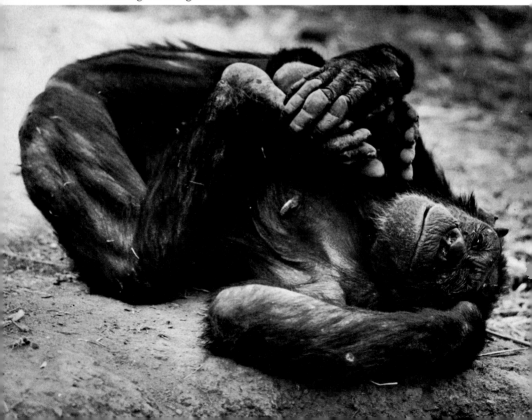

Flo became increasingly playful after Flint's birth (*Copyright National Geographic Society*)

Fifi sees her reflection
in Hugo's lens
(*Copyright National
Geographic Society*)

Flo

Someti▮
we hid ban▮
under our s▮

Flo grooming Flint

it seemed that Goliath accepted Mike's superiority, and a strangely intense relationship grew up between the two. They often greeted one another with much display of emotion, embracing or patting one another, kissing each other in the neck, after which they usually started grooming each other. During these grooming sessions it appeared that the tension between them was eased, soothed by the close, friendly physical contact. Afterwards they sometimes fed, or rested quite close to each other, looking peaceful and relaxed as though the bitter rivalry of the past had never been.

Indeed, it is one of the most striking aspects of chimpanzee society that creatures who can so quickly become roused to frenzies of excitement and aggression can, for the most part, maintain such relaxed and friendly relationships with each other. One day I followed Mike from camp, over the stream and for some way into the thick forest of the opposite mountain slope. With Mike were the old male J.B. and Flo with Flint, Fifi and Figan. Eventually they stopped under some trees in one of those places where chimps love to rest during the heat of the day. I sat down nearby. Fifi climbed high up into the tree and made herself a little nest: Figan and J.B. snoozed on the ground: Flo, with Flint sleeping on her lap, sat grooming Mike. After a while they too lay down to rest.

Presently Mike reached out towards Flo's hand and began, almost imperceptibly, to play with her fingers. Soon she responded, gently grasping his hand, twisting and pulling away – only to reach out and grasp it again. After a few minutes Mike sat up and leaned over Flo, tickling her neck and her so ticklish groin until, protecting Flint with one hand and parrying Mike with the other, Flo started to shake with panting gasps of chimpanzee laughter. After a while she could stand it no longer and rolled away from him. But she was roused, this ancient female with her stumps of teeth, and soon she was tickling Mike in the ribs with her bony fingers. Then it was Mike's turn to laugh and reach to grab her hands and tickle her again himself.

After ten minutes Flo, it seemed, could endure the tickling no longer and she moved away, leaving Mike stretched out with a benign expression on his face. And yet, just two hours earlier, this same male had attacked Flo savagely, dropping a huge pile of his own bananas and pounding the old female unmercifully – just because she had presumed to take a few fruits for herself from a nearby box. How was it possible for her to enjoy such a relaxed interaction with Mike so soon? The secret, perhaps, lies in the fact that, whilst a male chimpanzee is quick to threaten or attack a subordinate, he is usually equally quick to calm his victim with a touch, a pat on the back, an embrace of reassurance. And Flo, after Mike's vicious attack, and even while her hand dripped blood where she had scraped it against a rock, had hurried after Mike, screaming in her hoarse voice, until he had turned. Then as she approached him, crouched low in apprehension, he had patted her again and again on her head and, as she quietened, had given her a final reassurance by leaning forward to press his lips to her brow.

Would Mike have become the top-ranking male if I, and my paraffin cans, had never invaded the Gombe Stream? We shall never know, of course, but I suspect he would have, in the end. For Mike has a strong desire for dominance, a characteristic marked in some individuals and almost entirely lacking in others. Over and above which, Mike has unquestionable intelligence – and amazing courage, too. I shall never forget the time, soon after Mike had become the uneasy top-ranking male, when some of the other high-ranking males turned on him. Mike had charged into camp, hurled a few rocks, and, in passing, briefly pounded on David Greybeard. David Greybeard, in some ways, was a coward for he nearly always tried to avoid trouble and, when he couldn't, he usually tried to hide behind a higher-ranking companion, such as Goliath. But when he became really roused he could be a very dangerous chimpanzee.

On this occasion David, after running, screaming, away from Mike, turned and began to utter loud, fierce-sounding *waa* barks. He hurried over to Goliath and embraced him, then turned and again shouted towards Mike. By this time Hugo and I knew David well, and it was obvious that he was furious.

Suddenly David ran forward a short way towards Mike and, immediately, Goliath joined him, adding his own fierce call to that of his friend. Mike began to display, charging across the clearing towards another group of males. They fled, screaming, but then, as David and Goliath were still calling, they joined in too. Now it was five strong adult males, including the once top-ranking Goliath, against one. Again Mike charged across the clearing, and all at once, with David in the lead, the others were after him. Mike, screaming now, rushed up a tree, and the others followed. Hugo and I felt sure that this was the show-down: now Goliath would regain his lost position.

Suddenly, to our amazement, Mike turned – instead of leaping off into the next tree and running away, he turned. He was still screaming, but he began to sway branches violently and the next moment he took a leap towards the five. In a flurry of fright they rushed down the tree, almost falling over one another in their haste, and fled with Mike after them. When Mike sat, his hair on end, his eyes glaring, the others stayed away from him, cowed. Mike had won a spectacular victory by bluff.

*       *       *

When I refer to Mike as the dominant male, what I really mean is that he became top-ranking amongst those individuals that we know – individuals whose normal range includes our home valley. Once I had become really familiar with all the chimpanzees of our community, I quickly realised, from visits to the north and south of the Reserve, that there are, in fact, two other communities. Many of the individuals comprising these groups seldom or never travel as far as our centrally located valley, but there is, without doubt, some intermixing between chimpanzees of the three communities.

One fully mature male, whose normal range, so far as we know, lay to the south, did start to visit our feeding station; he would come for a week or so at a time when he was in the vicinity and then disappear back to his normal haunts. Just before he died he became quite a regular visitor to camp, but his relationships with the males of our group were always rather tense. Quite often females from the northern or southern communities arrive at camp during their periods of sexual swelling, brought along by our males: and once they have discovered bananas some of them become fairly regular visitors whilst others come only once or twice in a year.

On a number of occasions I have seen individuals from two of these main communities meet up and mingle without aggression, feeding together side by side. But it seems that Mike himself is reluctant to mix with the chimps to the north and south of his domain. A few times when 'strangers' called from a neighbouring valley, Mike, after much displaying and calling himself, turned back, taking some of his group with him whilst others moved on to mix with the strangers.

A chimpanzee community is an extremely complex social organisation: it was only when a large number of individuals began to visit the feeding area, so that I could make regular observations on their interactions one with another, that I began to appreciate just how complex it is. The members who comprise it move about in constantly changing associations and yet, though the society seems to be organised in such a casual manner, each individual knows his place in the social structure – knows his status in relation to any other chimpanzee he may chance upon during the day. Small wonder there is such a wide range of greeting gestures – and that most chimpanzees do greet each other when they meet after a separation. Figan, going up to an older male with a submissive pant-grunt, is probably affirming that he remembers quite well the little aggressive incident of two days before when he was thumped soundly on the back. 'I know you are dominant: I admit it: I remember,' is the sort of communication inherent in his submissive gesturing. 'I acknowledge your respect: I shall not attack you just now,' is implicit in the gentle patting movement of Mike's hand as he greets a submissive female.

As Hugo and I became increasingly familiar with Mike's community we began to learn more and more about the variety of relationships which existed between different adult chimpanzees. Some individuals only interacted when chance – such as a fruiting tree or a sexually attractive female – threw them together. Others moved about together frequently and showed an affectionate tolerance and regard for each other which, we felt, could best be described as friendship. And, as our study continued, we found that some friendships persisted over the years whilst others were of relatively short duration. We learnt, too, to appreciate the different characteristics of male and female chimpanzees. And the more we learned, the more we were impressed by the obvious parallels between some chimpanzee and some human relationships.

Firm friendships, like that between Goliath and David Greybeard, seem to be particularly prevalent amongst male chimpanzees. Mike and the irascible, testy old J.B. travelled about in the same group very frequently. When I first knew them, J.B. was the higher-ranking of the two, but Mike's strategies with the paraffin cans served to subordinate J.B. along with all the other males. However, once things had settled down, with Mike secure in the top-ranking position, it became apparent that J.B. had also risen in the social ladder. When he was in a group with Mike, J.B. was able to dominate Goliath as well as other males who had held a higher rank than he before Mike's rise. These other males quickly accepted J.B. as second to Mike, but Goliath asserted his old superiority over J.B. on many occasions when Mike was not part of the group. I well remember one day when Goliath threatened J.B. who had approached his box of bananas. J.B. at once moved away, but then began to scream loudly, looking across the valley in the direction which Mike had taken earlier. Mike must have been quite close because, within a few minutes, he appeared, his hair on end, looking round to see what had upset his friend. Then J.B. ran towards the box where Goliath sat, and Goliath, with submissive pant-grunts, hastened to vacate his place – even though Mike took no further active part in the dispute.

There was another occasion when, after eating about twenty bananas, J.B. tried hard to break open another box. As he was a past master at breaking boxes, and as they were hard to mend, Hugo and I tried to dissuade him from the idea by walking slowly to the box and sitting on it. J.B. did, indeed, move away, but then he climbed a nearby tree and screamed, again staring in the direction taken by Mike when he left earlier. That time, however (perhaps luckily for us), Mike must have been out of earshot for he did not come back.

Leakey and Mr Worzle were two other males who frequently travelled together. In temperament they were very different. Leakey, like his namesake, is robust, high-ranking and usually good-natured. Mr Worzle, on the other hand, was always nervous, both in his dealings with other chimps and with humans. He was very low-ranking indeed and, even before he became really decrepit prior to his death, was subordinate to all the other adult males – and some of the adolescent males also. Nevertheless, the two spent hours in each other's company, grooming each other, feeding and moving from place to place together, building their nests in the same or neighbouring trees. When Leakey was with him, Mr Worzle always seemed far more relaxed and confident.

Friendships of this sort are beneficial not only to the lower-ranking of the pair. One day, during the period when Goliath was losing his top-ranking position, he arrived in camp on his own. He was tense and obviously anxious about something; every so often he stood upright to stare back along the way whence he had come, and he jumped at every sudden sound.

All at once Hugo and I noticed three males, one of whom was the high-ranking Hugh, standing at the top of a slight rise and looking towards Goliath. They all had their hair slightly on end and, as they began running down the slope, they reminded us of a gang of thugs. Goliath did not wait to see what they would do; with great speed and very silently he ran in the opposite direction and vanished into the thick vegetation surrounding camp. The three rushed after him, and for the next five minutes they bustled about noisily in the undergrowth, obviously searching for Goliath. They were unsuccessful and presently they emerged and began to eat bananas. Suddenly Hugo pointed and there, a short way up the slope, I saw a head peeping cautiously from behind a tree-trunk – Goliath's. Every time one of the three looked up Goliath bobbed back behind his tree, only to peer out after a few moments. Presently we saw him moving off quietly up the slope.

The chimps slept near camp that night, and very early, almost before it was light, we heard a sudden burst of pant-hooting from the direction of Goliath's nest. Hugh and the other two males were the first to arrive in camp, dark shapes in the grey light of dawn. Then, as they were eating bananas, we heard a sudden burst of calling from up the slope. A moment later Goliath charged down, dragging a huge branch and hurling it forward as he crossed the clearing. Without pausing he rushed at Hugh and began to attack him. It was a fierce battle, and Hugh came off very much the worst: usually a male pounds his victim for a few seconds only, but this time the two combatants rolled over and over, grappling and hitting, and then Goliath managed to leap on to Hugh, hanging on to his shoulder hair and stamping on his back with both feet.

It was just after the start of the fight that Hugo and I realised why Goliath was suddenly brave; we heard the deep, characteristic pant-hoots of David Greybeard, and glimpsed him charging in his slow and magnificent fashion across the clearing and past the battling males. David must have joined his friend early that morning and, by his presence alone, given Goliath the courage to face Hugh and his gang.

With the exception of David and Goliath, who bore no resemblance at all to each other, we have been able to detect similarities in either physical make-up or behavioural characteristics – or both – in all of the pairs of male friends that we have known. This was particularly striking in the case of Leakey and Mr Worzle. Mr Worzle had extraordinary eyes, for the part around the iris was white instead of being heavily pigmented with brown as in other chimpanzees. His eyes, therefore, exactly resembled those of a man. Leakey, too, showed the same unusual lack of pigmentation, though to a much lesser extent than Mr Worzle. We suspect, in fact, that pairs of male friends may often be siblings.

The only two adult females we know who enjoyed this sort of friendship were almost certainly sisters; not only did they look alike facially, but they had the same massive build, and both were prone to perform charging displays, stamping on the ground and swaggering in a manner more typical of males. They were the only two adult females I ever saw playing with each other, rolling about on the ground, tickling one another and panting with laughter, each with her infant cradled in one arm.

The adult females of the chimpanzee community are almost always submissive to adult males – and, indeed, to many of the older adolescent males. But they have their own dominance hierarchy of which Flo, for many years, was supreme, respected and even feared by old and young females alike. Flo was exceptionally aggressive towards her own sex, and she would tolerate no insubordination from young adolescent males. Much of her confidence no doubt resulted from the fact that she was so often accompanied by her two eldest sons and, with the aggressive Fifi as well, the family was formidable indeed.

As mentioned earlier, Flo at one time often wandered about together with the mother, Olly. But their relationship was very different from that between Mike and J.B. or David and Goliath. For one thing Flo was frequently aggressive towards Olly, and for another, neither would go to the assistance of the other in times of trouble. The only time I did see them united was when they ganged up on a young stranger female.

This female had first appeared in camp surrounded by a retinue of males and boasting a pink swelling. She had come in every day for ten days and become quite used to the strange place where bananas grew in boxes on the ground. Often she had been in camp at the same time as Flo and Olly, but the two older females had seemed to ignore her completely.

And then one day, when a small group of chimps including Flo and Olly were in camp, Hugo and I saw the young stranger female. Her swelling had quite gone; she was sitting in a tree at the edge of the clearing looking nervously towards us. We were pleased for, at that time, only a very few young females visited the feeding area. Just as we were getting some bananas ready for her we noticed Flo and Olly staring fixedly at the stranger, every hair on their bodies bristling.

It was Flo who took the first step forward, and Olly followed. They went quietly and slowly towards the tree, and their victim failed to notice them until they were quite close. Then, with pants and squeaks of fear, she climbed higher in the branches. Flo and Olly stood for a moment, looking up, and then Flo shot up the tree, seized the branch to which the now screaming female was clinging and, her lips bunched in fury, shook it violently with both hands. Soon the youngster, half shaken, half leaping, scrambled into a neighbouring tree with Flo hot on her heels and Olly uttering loud *waa* barks on the ground below. The chase went on until Flo forced the female to the ground, caught up with her, slammed down on her with both fists and then, stamping her feet and slapping the ground with her hands, she chased her victim from the vicinity, Olly still barking, ran along behind.

When the stranger had vanished along the forest track Flo stopped. Her face was spattered with liquid dung, product of the young female's terror, her hair was still bristling. Olly stood beside her and the two old females listened as the sounds of screaming gradually receded up the valley. Then Flo turned, wiped off the dung with a handful of leaves, and slowly returned to camp – where, throughout the incident, Fifi had been looking after the eight-month-old Flint.

This was not the only occasion when we witnessed sudden alliances between two or more adult females which resulted in driving young newcomers, of the same sex, away from the feeding area. We have not, however, seen them gang up in this way on

A group of chimpanzees resting in a tree

(*Overleaf*) It was a vast, rugged country in which to search for chimpanzees

stranger adolescent males: nor have we seen adult males of our group driving away strangers of either sex from the feeding area. What, then, motivates the aggressive behaviour of these females? Is it perhaps the fact that older females, who normally have a much smaller range than males are more territorial? Or could it be due to some more complex emotion – do old females, perhaps, resent the attention paid to young stranger females by 'their' adult males? Are they, in other words, motivated by the emotion which, in human beings, we call jealousy? We cannot be sure – but sometimes it certainly seems like it.

One day, when Flo was socially grooming with four adult males, a young pregnant female arrived; she had recently joined our group from the north. Pregnant females often continue to show monthly swellings, and this one had a very pink posterior. The males, on this occasion, did not mate her, but they were, nevertheless, interested: they left Flo, hastened over to the newcomer, inspected her bottom, and began to groom her vigorously. It was only a couple of minutes later that I noticed Flo. She had moved a few yards towards the young female and was standing staring at her with every hair on end. Had she dared she would, without doubt, have attacked the newcomer. As it was, she presently walked slowly over to the group and herself inspected the swelling carefully. Then she moved away and sat down to groom Flint.

We could scarcely believe it when, the following day, Flo showed the beginnings of a swelling. Flint was under two years old and, whereas young females may start swelling again when their infants are only fourteen months, old females like Flo do not normally become pink again for four or five years after giving birth. However, Flo's sex skin was swollen enough to arouse instant attention from Rodolf who feverishly pushed her to her feet and intently inspected her bottom. So did a couple of other males. Then they sat around grooming her. The next day that extraordinary swelling had gone – nor did Flo show any signs of swelling again for the next four years. I cannot believe that it was pure coincidence.

The female chimpanzee is, indeed, very different from the male, although, as in the case of humans, some females show masculine characteristics, and vice versa. Adult females, typically, resort to pleading with many of the gestures and calls made by infants when they are trying to get their own way with a social superior.

121

Melissa, when begging from a male, reaches out her hand time and again, touches him ingratiatingly and, if this behaviour fails, may start to whimper or even scream like a child in a tantrum. Like other females, she can be very persistent in her begging so that, often, she is eventually rewarded with a scrap of banana or cardboard or whatever it is she wants. Once, when Mr McGregor was grooming her, he turned to groom a higher-ranking male who had joined the group. Melissa stared at his back for a while and then began rocking back and forth and whimpering. He ignored her. Her whimpers grew louder and, every so often, she reached out and gave him a quick poke with her finger. Still Mr McGregor continued to groom the other male. Finally, almost screaming in her frustration, Melissa reached out and gave him a hard shove with one foot – and at this the old male finally turned and began to groom the importunate female again.

It appears that females are more likely than males to harbour grudges. At one time Melissa, if she was threatened by a superior, nearly always hurried over to a higher-ranking individual and, whilst reaching out to touch him, directed loud screams towards her aggressor. Obviously she was trying to incite her chosen champion to retaliate on her behalf. The fact that the males she approached seldom responded, except, perhaps, to reassure the noisy female, in no way dampened her ardour; she did exactly the same the next time she was threatened. One day, when she was lightly cuffed by Rodolf, it happened that he was the highest-ranking male in the group. But to our astonishment when Mike arrived, some ten minutes later, Melissa rushed up to him, pressed her mouth to his neck and then, with one hand on Mike's back, she started screaming whilst staring at Rodolf and making little flapping movements towards him with her other hand. As usual, her strategem was ignored, but we saw her behaving in exactly the same way on other occasions.

Melissa was by no means the only female to nurse a grievance over a length of time. Pooch, almost certainly, lost her mother when she was between five and six years old, and she struck up a strange relationship with an old male, Huxley. For the most part they paid little attention to each other, although sometimes they sat grooming together; but whenever Huxley got up to leave, Pooch followed like a shadow.

One day, when a large group of chimps had visited camp, Pooch, who was about six years old at the time, stayed behind with Evered, a year her senior, when the others left. Neither had managed to get any bananas. As soon as the group was out of sight we gave them some fruit; a squabble broke out and Evered cuffed Pooch who screamed. Then she turned her rump, presenting submissively, he patted her, and they sat peacefully side by side feeding. We were astounded when Pooch, a few minutes later, suddenly dropped her bananas and attacked Evered, biting him and pulling at his hair. Evered was probably startled too, for it is unusual indeed for a female to attack a male older than herself.

Then we realised what had prompted Pooch's behaviour, for we saw old Huxley, his hair on end, standing a short way along the track and staring towards us. His glance moved from Hugo and me to the youngsters – probably he had associated Pooch's earlier scream with us and, for that reason, hurried back to her rescue. A moment longer he stood and then charged down towards the squabbling chimps. It looked as though he cuffed them both; then he turned and plodded away again. Evered screamed until he got cramps in his throat and sat doubled up, as though in pain. Pooch immediately started after her protector and, as she passed Evered, she glanced at him with an expression I have never seen before or since in a chimpanzee. It looked exactly like the smirk a little human girl might be expected to give under similar circumstances.

What Pooch did to Fifi, a couple of years her junior and a play-mate of long-standing, was completely against chimp etiquette. It happened when the two were romping together and Fifi, probably accidentally, hurt Pooch who screamed and hit at Fifi. Fifi grinned in fear and then turned her rump submissively and presented to Pooch. Pooch should have reached out and touched Fifi's bottom: instead she leaned forward and, deliberately and rather hard, bit Fifi's little pointed clitoris.

Fifi, with all her mother's staunchness of spirit, turned and flew at the larger female. The two rolled and grappled on the ground, pulling out handfuls of each other's hair until Flo finally arrived, her hair on end, and Pooch retreated, screaming. Fifi, her screeches cramping in her throat, presented again – this time to her mother, and Flo reassured her, patting her rump again and again until the child quietened. But her bottom swelled up and bled a good deal.

Obviously it was very painful, and she made herself a soft leafy nest on the ground where she reclined for a while, gently dabbing at her wound with a handful of leaves.

There is, indeed, a great deal in chimpanzee social relationships to remind us of some of our own behaviour; more, perhaps, than many of us would care to admit. Only by carrying on our research for years to come, and studying the social structure in a group where blood-relationships between the different individuals are known, shall we succeed in understanding the whole complex and intricate pattern.

Chapter 11  The Growth of the
Research Centre

When first I set foot on the sandy beach of the Gombe Stream
Reserve, I never dreamed that I was taking the first step towards
the establishment of the Gombe Stream Research Centre: that,
nine years later, there would be ten or more students studying
not only different aspects of chimpanzee behaviour, but also
observing baboons and red colobus monkeys.

We took on our first research assistant, Edna Koning, soon after
Flint's birth. Edna wrote begging us to take her on in some capacity
and, since there had been too much work for me alone even before
Flo produced a son, we felt it was an excellent idea.

Edna not only typed out my notes but also learned to make
accurate observations herself. Then, when I followed Flo and Flint
into the mountains, I knew that if anything exciting happened in
camp Edna would record the events.

In those days we worked all day and far into the night. I dictated
my observations on to tape which meant that I didn't have to take
my eyes off the activity around me. Edna typed out the tapes in the
evening whilst I struggled with analysis for my PhD thesis. We
started making an extra copy of the notes – three copies in all – and
I marked this copy into categories of behaviour – grooming, submission, aggression, and so forth. Edna, Hugo and I cut these up
and pasted them, in their relevant sections, into large files. This, of
course, was immensely helpful for my analysis. The third copy was
always sent off, every month, to Louis for safe keeping – in case of
fire, floods or some other catastrophe at the Gombe.

Then, too, there was dung-swirling. It was Hugo who first
thought of washing the chimps' dung rather than examining dried
samples for indications of what food had been eaten. Chimpanzees
swallow many stones of fruits when they are feeding, so that we
always had a good idea of what species were currently ripe. It is

amazing how much of a chimpanzee's food seems to pass through the digestive tract only partially digested. Dung-swirling was an excellent method of finding out how often members of our group ate insects and meat, and data gathered in this way, together with information gleaned from watching chimpanzees feeding, gave us a very good picture of their diet throughout the year. We swirled the dung down by the stream, putting each specimen in a large tin with holes in the bottom, adding water, and swirling it round and round over a hole we had dug.

Hugo, in addition to helping me with my analysis, had a lot of his own work. He was still, of course, financed by the National Geographic Society, and he did all the accounts, both his own and mine. He had to write out a lot of legends for his film and 35 mm photos and, particularly in the rainy season, he had a constant battle to keep his photographic equipment in working order.

We all worked so hard that Vanne, who came to visit us for a while that year and who, of course, was instantly roped in for her share of the chores, suggested we take off one evening a week from our toils. This was an excellent move, and we looked forward to our rest nights as many people look forward to the week-ends. On these precious evenings we played taped music, had a couple of drinks round a fire, and enjoyed a leisurely supper – instead of swallowing down our food in one of the tents before rushing back to work. Sometimes, too, we had hilarious sessions of liar dice.

Nonetheless, even on these evenings, our conversation was almost entirely 'chimp'; if our work had not also been our pleasure it is doubtful whether we should have been able to keep up the pace. We were all totally immersed in the goings on of our chimpanzee group – as Hugo so often said, it was like being spectators of life in some little village. Endless fascination, endless enjoyment, endless work.

Towards the end of that year, try as we would, we simply couldn't keep up with all the work – for new chimpanzees were continually joining the regulars at the feeding area, and there was Melissa's baby, Goblin, as well as Flint demanding very detailed observation. So we took on a secretary, Sonia Ivey, and she, like Edna, became fascinated with the work and with the chimpanzees. Sonia typed out all my tapes, which gave Edna more time to make observations and I to pursue my investigations in the forests. The chimps became increasingly tolerant of my following them closely so that I was able to stay with them for longer and longer at a stretch.

By this time we had a group of forty-five chimpanzees visiting camp, some of them regularly, like the Flo family, and others, from groups normally ranging to the south or north, only when their wanderings brought them close to our valley. Apart from some of the very infrequent visitors, these chimps were so used to the set-up at the feeding station that they showed no hesitation in wandering in and out of our tents, taking anything they fancied.

We had, of course, profited from Kris Pirozynski's experiences: we were careful to put away all clothing in tin trunks – including our bedclothes which we had laboriously to fold up each morning. Once, just after dawn, I heard an anguished yell from Vanne. When I went to investigate I found that David Greybeard had caught her dressing. She was sitting half naked on her bed clinging frantically to one pyjama leg whilst David sat beside her, one hand on her knee, sucking the end of the other leg. When I could stop laughing I fetched a banana and stood with it outside the tent. David, though he pulled his part of the pyjamas quite hard, eventually let go and accepted the banana as a substitute, whilst Vanne quickly zipped up the tent and bundled her pyjamas to safety.

There was no one to offer Rodolf a banana when, way up in the mountains, he suddenly left off grooming his chimpanzee companions and, approaching me with his hair on end, seized hold of my shirt and pulled. He looked so ferocious that I was just about to strip when, with his hair slowly sinking down, he sat close to me and began to suck the cloth where he was. He stayed there fifteen minutes and then, with a tiny corner he had torn off, returned to his grooming.

Indeed, after a while, we began to feel that nothing was sacred. Quite often we hid a couple of bananas in the pockets of our clothes, ready to smuggle to Fifi or some other youngster who had been unable to get a share from the boxes. But, as the adult males quickly realised what was going on, we had to stop. The chimps, however, remembered the old hiding-place for a long time. One morning Hugo, just dressed and half asleep, let out a yell when Leakey appeared in the tent doorway, stared hard at him, lifted up his shirt and poked an inquiring finger right into his navel! It was Leakey, also, who noticed what seemed to him to be a most promising bulge in Edna's shirt, reached up, and gently squeezed her breast.

One day, unthinkingly, I put a tasty banana into my pocket for my own consumption. When Fifi tried to push her hand in I moved away. Fifi stared very hard at the bump in my pocket and then picked a long thin grass which she very gently inserted deep down into my pocket. She withdrew it and intently sniffed the end: this, apparently, told her all she wanted to know, and she followed me round, whimpering, until I finally had to give her the fruit.

We also had to keep our egg supply very carefully hidden. Mr McGregor, Mr Worzle and Flo were the three chimps who loved eggs the most. One day old McGregor managed to make off with four hard-boiled eggs which Edna had prepared for lunch – and it was well worth the loss for the laugh he gave us. The chimpanzee almost always eats an egg together with a large mouthful of leaves; only when sufficient leaves have been stuffed into his mouth, along with an egg, does he crack the shell. Then he sits savouring the egg-leaf wadge for minutes on end.

Mr McGregor looked startled when he put the first of the eggs into his mouth – no wonder, for it was hot. He took it out, looked at it carefully, sniffed it, and then shoved it back in with copious handfuls of leaves. Then we heard him crack the shell. This obviously was even more puzzling – no delicious liquids ran into his mouth. He spat out the whole mass of leaves and egg and stared at it. He tried all four eggs, each time stuffing in more and more fresh leaves until he was surrounded by fragments of white and yellow egg and mounds of crumpled greenery.

Another of our problems that year was to protect the tents. We fixed all the guy ropes high up on to trees, or on to a thick wooden railing which Hassan built around the tent – for the chimps had found that to yank out tent pegs, one by one, as they charged past during an arrival display, produced spectacular results. For a while we had no further trouble. Then one day Goliath displayed straight through our open tent and, as he did so, snapped in two each of the thick wooden tent poles, one after the other, as though they had been matchsticks. In his wake he left a crumpled mass of canvas partially held up by those guy ropes that were attached to trees.

Charging displays can help a male's rise to dominance

Mike bows to the dominant Goliath

Mike uses noisy paraffin cans to help him take over as dominant male
(See p. 109) (*Copyright National Geographic Society*)

The dominant Goliath reassures Figan by touching him. Figan gradually relaxes

Finally we got some sturdy tree-trunks and, with Hassan's help erected these, sunk in cement. They made somewhat unconventional but very satisfactory tent poles.

For most of that year the chimps' banana supply was our worst headache. For one thing there never seemed to be enough boxes: Hassan was making more almost non-stop, but every day two or three – or even more – were put out of action by one of the chimps. Even when everything had been sunk into cement, some of the adult males still managed to break something. J.B. was the worst offender. He kept breaking off the steel handles of the levers, so that we could not close them. And he managed to snap even strong cable – though the only part showing was a length of about seven inches between the cemented end of the pipe and where it was attached to the lever. It was a rather terrifying indication of the superhuman strength of these chimpanzees.

It was all right if a group came in and left before the next group arrived – then we had a chance to re-fill the boxes from our little store. But we dared not open the store with a group of chimps in the vicinity. As it was we were often surprised during an attempt at replenishing the boxes; and if J.B. or Goliath or one of the other big males swaggered up when we were holding a bucket full of bananas we simply had to give them the lot – they were too powerful to trifle with.

Worst of all was the problem of David Greybeard's banana supply. David remembered the good old days when he, Goliath and William had been able to make off with any bananas that were available. In those days he didn't have to compete for a share with five to ten other hungry males. And David never hurried when he arrived at camp: he liked to let the other males rush on ahead so that by the time he arrived most of the excitement would be over. Yet somehow, when he did come, we had to have bananas for him; otherwise, with his lower lip pushed determinedly forward, he plodded into one tent after the other looking for them. He pulled everything apart, pushed everything over – even ripped open mosquito net windows if the tent was zipped shut. The problem was, of course, that the other chimps also looked through the tents for hidden fruit – though not so thoroughly and disastrously as David. They quickly learned our different hiding-places, so we constantly had to think of new ones.

Even when David's supply remained intact we still had the difficult task of getting it to him without the other more aggressive
males noticing. And after we had, with great trouble, succeeded in
getting a pile right into his arms, then Flo or Melissa or some of
the other females usually clustered round and reached out, one
after the other, to take fruits from David's pile. David seldom
objected – after all, *he* would always be able to get more! Life became
increasingly hectic and impossible and, more and more, I longed
for those uncomplicated days when I roamed the mountains.

In 1965, however, things began to look up. The National Geographic Society, which was still supporting the research, granted
us funds for the erection of some prefabricated aluminium buildings. We chose another site, still farther up the valley, which offered
a superb view of the mountain opposite and a glimpse of the lake.
Once again all the work had to be done at night but, apart from
the preparation of the cement floors, it did not take long to set up
the buildings. When they were ready and covered on the sides and
roofs with grass, they blended in well with their forested surround-
ings. The largest one contained a fairly big room for working and
the filing of records, two very small rooms to serve as bedrooms
for Edna and Sonia, and two even smaller ones as a kitchen and
store. The other building was for Hugo and myself. In addition, we
put up a tiny hut as a banana store down in the valley just
below the new camp.

This time we had even less difficulty in acquainting the chimps
with the new site. One morning, when Hugo and I were up at the
completed buildings checking that everything was ready, we
looked over the valley and saw David Greybeard and Goliath
feeding in a palm tree. What luck! Quickly we held up a large
bunch of bananas. The two males screamed and hugged each other
for fully a minute before rushing down and hastening across the
valley. As it happened, fifteen other chimps were scattered about
the valley and, upon hearing the exciting calling of David and
Goliath, they too hurried towards us; the whole group converged
upon the new feeding area. The only disappointment was that,
since the attempt was unplanned, Hugo had no camera to record
the wild excitement, the embracing and kissing and patting and
screaming and barking that went on before the chimps had calmed
down sufficiently to start feeding.

Within three days nearly all the chimps – all but the most ir-regular visitors – had discovered the new site and we were able to close down our other feeding area completely. The new buildings were so luxurious compared to anything Hugo and I had ever known at Gombe that it was heartbreaking for us to have to leave the place a few weeks after everything was finished. But Hugo had to start another photographic assignment – for the National Geo-graphic Society could not afford to maintain a photographer full-time at the chimpanzee camp – and I had to spend nine months in England to finish writing up my results for my PhD dissertation.

Not until Hugo and I had actually left the Gombe Stream did we realise that, during the year, we had made one grave mistake. We had encouraged Flint to touch us and we had tickled him gently. It had been a delightful experience, and Flint had become more and more trusting: we had marvelled that a wild chimpanzee mother could lose her fear of humans to the extent of allowing her infant to play with us. But Fifi had copied Flint's example, and so had Figan. At the time it had not seemed to matter; it proved, as my grooming of David had proved, that it is possible to establish a close and friendly relationship with a creature who has lived the first years of his life in fear of man. Hugo and I were actually able to tickle Figan, to wrestle with him, rolling on the ground, al-though, even when eight years old, the young male was much stronger than either of us.

But when we left, when the potential of the new research build-ings dawned on us, when we started receiving letters from students asking if they could join our team, we realised the foolishness of our behaviour. For one thing the adult male chimpanzee is at least three times stronger than a man; if Figan grew up and realised how much weaker humans really were, he would become danger-ous. Moreover, repeated contact with a wild animal is bound to affect its behaviour. We made a rule that, in future, no student should purposefully make contact with any of the chimpan-zees.

Edna and Sonia – who had soon learned to make accurate obser-vations herself – kept things going for almost a year on their own. Since then we have had a constant stream of research assistants working at Gombe. Gradually our research programme has ex-panded so that, for the past few years, we have offered facilities to those wishing to study baboon and red colobus behaviour. Some students have worked at the Gombe for one year as my research assistants, keeping up the all-important general record of the behaviour of known chimpanzees; most of these young people had B.A. degrees, and all put in hours and hours of work and added considerably to our knowledge of chimpanzees. Some of them elected to stay on for a second year working as research students on some chosen aspect of chimpanzee behaviour.

In 1967 there was a major change in the status of the Gombe Stream: it was taken over by Tanzania National Parks and became the Gombe National Park. Game Department scouts gave place to National Parks rangers with their own quarters down in the south of the Park. Together with the National Parks authorities, we are slowly working our way towards opening a second feeding station, in the south, for the benefit of interested tourists and visitors. For two years now students have been working in the proposed area, living on their own, trying to habituate the chimpanzees of the southern group as, in 1960, I struggled to get our own group used to me.

And so the Gombe Stream Research Centre has gradually expanded. To-day there are eight little sleeping houses up at the observation area, nestling out of sight in the surrounding trees. There are three larger buildings down on the beach, and three more huts for the students working on baboons and monkeys. The African staff alone have quite a village down on the beach grouped around the house of old Iddi Matata. Dominic and Sadiki and Rashidi are all with us still – and there are many more. It cannot be said that conditions at the Gombe Stream Research Centre are, in any sense, luxurious. But the facilities are more than adequate for young people who love animals, exciting and fascinating research – and hard work.

The biggest challenge which we have faced, at any time, has been the actual presentation of the bananas: how to offer them in a way most similar to a natural food supply, and so as to affect as little as possible the social behaviour of the chimpanzees. There has been much trial and error throughout the years.

To start with, of course, we gave the chimps bananas whenever they came into camp. It was so exciting to Hugo and me to be able to film and watch individuals at close quarters and regularly, that we didn't worry too much if they visited the valley more frequently than they would have done had the feeding station not existed. Also, in those early days, the idea of really long-term research had not been born, and we were anxious to record as much as we could before we had to leave for ever.

After a few years, however, we realised that the constant feeding was having a marked effect on the behaviour of the chimps. They were beginning to move about in large groups more than they had ever done in the old days. They were sleeping near camp and arriving in noisy hordes first thing in the morning. Worst of all, the adult males were becoming increasingly aggressive. When we first offered the chimps bananas the males seldom fought over their food: they shared boxes or, at worst, chased off another individual, or threatened, without actually attacking.

When Hugo and I returned to the Gombe in 1966 after I had successfully completed my terms at Cambridge, we were horrified at the change we saw in the chimps' behaviour. Not only was there a great deal more fighting than ever before, but many of the chimps were hanging around in camp for hours and hours every day. This was almost entirely due to Fifi and Figan and, to a lesser extent, Evered.

These three youngsters had discovered that, in order to open a banana box, all they had to do was to pull out the simple pin that served to keep the lever closed. Laboriously Hassan had worked at the fastening, cutting threads into both pin and handle so that it had to be unscrewed rather than simply pulled out. This had worked for a couple of months, but eventually the same three youngsters had solved that problem too. Then Hassan had fixed nuts on to the ends of the screws, so that those had to be removed before the screws could be unscrewed. Just before Hugo and I got back, Figan, Fifi and Evered had mastered that. Things were chaotic.

Evered would go up to a handle, unscrew it, and then, with loud food barks, hurry to the box which he had thus opened. So, of course, did any other chimps in the vicinity, and it was rare for Evered to get more than a couple of bananas, if that, unless he happened to be the only chimp, or the highest-ranking one, in camp. Usually he simply went round opening box after box until, eventually, he had fed all the others – and then, if there was a box left, he got that himself. His only attempt at guile was to arrive earlier and earlier in the morning – presumably in an endeavour to be first in camp and have the field to himself. But the others came earlier and earlier too!

Fifi and Figan were far more astute. Both of them learned quickly that, however many boxes they opened, they were unlikely to get any bananas when there were higher-ranking chimpanzees present. So they just lay around, together with Flo, waiting for the others to go. Then, when there were no adult males in camp, they would quickly open a box each. Sometimes they could not resist going up to a handle and unscrewing the screw. But they did not then release the lever and hurry to the open box: they just sat, with one foot keeping the lever closed, casually grooming themselves and looking anywhere except at the box. Once I timed Figan sitting thus for over half an hour. But, of course, though the other chimps had not mastered the secret of the screws, they were bright enough to realise that, if they hung around, Fifi or Figan might, eventually, provide them with bananas. So they stayed in camp longer and longer. Sometimes they did manage to wear out Fifi's patience – particularly Mike, but not often. And so, day in day out, the Flo family remained in camp – and it amused us to see that it was Flo who occasionally tried to lead her offspring away just as, the year before, they had tried to lead her away from termite heaps. Only Flo had no lure and, after plodding off down the path, looking back over her shoulder again and again, she nearly always returned and collapsed, once more, in the shade of a palm tree.

Whilst we were impressed by the intelligence shown by Figan and Fifi, it meant that we had to devise a completely new system of banana feeding. Our next attempt was to install a large number of steel boxes, made in Nairobi. These were battery-operated and could be opened by pressing buttons inside the research building. One advantage of this system was that, when a large chimpanzee group arrived, we could offer each of the adult males his share at more or less the same time; they no longer had to hang around waiting and becoming increasingly aggressive. Also the chimps began to associate bananas more with boxes than people, for most of them never connected the pressing of a button with the opening of a box.

We decided, in addition, to feed bananas on an irregular schedule

– for two or three days running there would be no bananas at all. We hoped, in this way, to prevent the chimps remaining in our valley for days at a time. The system worked well during 1967, but we had by no means achieved the final answer. ·

The next problem was one that had been ever-present but which had been growing steadily worse each year: competition at the feeding area between chimpanzees and baboons. In 1968 when Hugo and I again returned to the Gombe for a few months, we found that things were in chaos. One baboon troop – the 'Camp Troop' as it was known, had taken to hanging around the feeding station, either in the nearby trees or on the other side of the valley from where there was a good view of our buildings. As soon as a chimpanzee group arrived, the baboons rushed in to try to get a share in the bananas. Over and above this, a second troop – 'Beach Troop' – had started spending several hours each day in the vicinity of camp as well.

The adult male baboons had become very aggressive, not only to the chimpanzees, but to humans as well. Some of our students were worried by the situation, particularly the girls – and rightly so, for a male baboon can be quite as dangerous an adversary as a leopard.

For a while we tried to discourage the baboons by refusing to open laden boxes whilst they were near camp, but this merely built up tremendous tensions and frustrations in the chimpanzees who were there. Baboons and chimps alike knew that there were bananas in the closed boxes and the longer we delayed opening the worse the situation became. Aggressive interactions multiplied and, when the boxes were finally opened, there was bedlam. Something had to be done, and done quickly.

First of all we stopped feeding altogether. As Hugo and I had anticipated, the chimps, when they arrived day after day and found all the boxes open and empty, began visiting camp less and less frequently. Within a week it was very quiet, with small groups occasionally wandering through, peering at the boxes, and trailing off again. The baboons stopped waiting around also.

After three weeks we began feeding again, very irregularly, filling the boxes only when it was certain that the baboons were sleeping far away and would not be around early in the morning. Meanwhile, work was going on for the construction of an underground bunker, stretching some ten yards from the main observation building.

The finished bunker is about four feet wide and high enough for a person to walk upright. There is plenty of room for storing bananas inside, and boxes dug up from the slopes have been placed along either side. At long last we have complete control of when and whom we feed. If some boxes have been filled and then, before the chimps have eaten, baboons come in, someone goes down into the bunker, removes the fruit, and closes the back of each box. Then we open the front of the boxes – and, so far as chimps and baboons are concerned, there are no bananas that day. If a small group of chimps arrives it can be fed, with the appropriate number of boxes and bananas, with no difficulty at all.

Since the construction of the bunker we have had almost no problems – only, when the back of one box broke, Goblin discovered that he could squeeze through into the tunnel. He would emerge from the box with an armful of bananas purloined from the supply within.

It is now possible to regulate quite carefully the frequency with which individual chimps are fed, and we ensure that none has bananas more than once in ten days or two weeks. So the chimpanzees have to a large extent resumed their old nomadic habits, and wander through camp, for the most part, only when they happen to be in the vicinity. This means, of course, that we cannot get as much information as when they all came more frequently, but they pass through sufficiently often for the students to make fairly regular records on many individuals.

For the past few years we have collected information on the chimpanzees' behaviour both at the feeding area – where it is possible to learn a great deal about changes in dominance and other relationships between individuals as well as to collect a wealth of data on infant development – and in the forests, under more normal conditions. The chimpanzees themselves have become amazingly tolerant of humans wandering along the mountain tracks behind them, almost part of their group. And, if they do get irritated, it is the easiest thing in the world for them to shake off their followers in that rugged terrain.

How long the present feeding system will continue to operate smoothly we cannot say – at present it seems as though, at long last, we have hit upon the ultimate solution. When I look back through the records on the different individuals, the quickly accumulating life-histories of chimps such as Flint and little Goblin, how worth while have been the struggles and heartaches and near despair.

*Chapter 12*  # The Infant

The birth of a baby is something of an event in many animal and human societies. In the chimpanzee community, where mothers only have an infant once every four or five years, births are relatively few and far between – not more than one or two a year in our group of thirty to forty individuals. So the appearance of a mother with a brand new baby often stimulates much interest amongst the other chimps.

I shall never forget the day when Goblin was first introduced to a large group. He was two days old, he could only grip on for a couple of steps without support from his mother, Melissa, and he was still attached, by the umbilical cord, to the placenta. The group was peacefully grooming in a tree, but as Melissa approached and began to climb I sensed the tension as first one and then another chimpanzee stared towards the new mother. Fifi instantly swung over towards her. Melissa, moving carefully, went to greet Mike, pant-grunting submissively and reaching to touch his side. She presented and he patted her rump, but when he moved forwards, staring at Goblin, she hastily moved away. It was the same when she greeted Goliath: he too wanted to see the baby more closely. So did David and so did Rodolf.

After five minutes Mike began to display, leaping through the tree and swaying the branches; Melissa, screaming, jumped away from him. As she did so, the placenta almost caught in some twigs. I was suddenly scared that the newborn might be torn from his mother's breast, but Melissa gathered up the cord just as Goliath displayed towards her. Soon the entire group was in confusion, with all the adult males leaping around and swaying branches, and not only Melissa but also the other females and youngsters rushing out of the way and screaming. It looked for all the world like a wild greeting ceremony for the new baby, though in reality it was un-doubtedly provoked by a sense of frustrated curiosity – Melissa

138

simply would not let the males get close enough to examine the baby properly.

Eventually things calmed down and the males began to groom each other again. Then Fifi and the other young females in the group gathered around the mother and baby to stare and stare at Goblin. If they got too close Melissa threatened them with a soft bark or upraised arm, but she did not move away and they were able to look their fill.

Since that day we have seen similar displays when new mothers first appeared with their babies; but, normally, only when the mothers themselves were young. There is usually far less commotion when an old female, such as Flo, appears with yet another infant. Since the experienced female does not run away from them the other chimpanzees are able to satisfy their curiosity. We have seen a group of four males sitting calmly very close to an old mother, staring intently at her newborn. When a mother does run away the situation is potentially very dangerous for a newborn baby, since, for the first few days of its life, the chimpanzee infant often does not seem able to grip well to its mother's hair. Usually, too, there is the added danger of a dangling placenta, for, so far as we know, the chimpanzee mother, in the wild, makes no attempt to break through the umbilical cord herself. We have never actually seen an infant dropped or hurt during these wild displays; on the other hand, we know of several babies which have mysteriously disappeared during the first few days of life.

During the six years since Flint and Goblin were born, twelve healthy infants have been born to our group and, although some of them died before they were a year old, our observations on them and their mothers have taught us much. Babies of under five months are normally protected by their mothers from all contact with other chimpanzees except their own siblings. True infants, from the age of three months onwards, often reach out to other chimps sitting close by, but usually their mothers pull their hands quickly away. It was, however, very different for Pom, one of the first female infants born into our group. Her mother, Passion, actually laid the baby on the ground, on the very first day of her life, and allowed a couple of young females to touch and even groom her as she lay there. But then, in all respects, Passion was a somewhat unnatural mother.

She was no youngster, this Passion. She had been fully mature when I first got to know her back in 1961, and I know she lost one infant before Pom's birth in 1965. Indeed, if her treatment of Pom was anything to go by, I suspect Passion had lost other infants too, for Pom, right from the start, had to fight for her survival. When she was a mere two months old she started to ride on her mother's back – three or four months earlier than other infants. It started when Pom hurt her foot badly. She could not grip properly and Passion, rather than constantly support the infant with one hand as most mothers would have done, probably pushed Pom up on to her back. The very first day that Pom adopted this new riding position, Passion hurried for about thirty yards in order to greet a group of adult males – seemingly quite without concern for her infant. Pom, clinging on frantically, managed to stay aboard – though much older infants, when they start riding on their mother's back, usually slide down if their mothers make sudden movements.

We expected that, when her foot was better, Pom would revert to clinging underneath Passion. But nothing of the sort happened. It is probably more comfortable for the mother when the child rides on her back – and having achieved this happy state of affairs . three months early, Passion had no intention of letting things change. Even when it poured with rain and Pom whimpered as she tried to wriggle under her mother's warm body for protection, Passion seldom relented but, again and again, pushed her infant back on top.

Most of the mother chimpanzees we have watched were helpful when their small babies nuzzled about, searching for a nipple. Flo had normally supported Flint so that suckling was easy for him throughout a feed – even when he was six months old. Melissa too had tried, though often she had bungled things and held Goblin too high so that his searching lips nuzzled through the hair of her shoulders or neck. But Passion usually ignored Pom's whimpers completely; if she couldn't find the nipple by herself it was just bad luck. If Pom happened to be suckling when Passion wanted to move off, she seldom waited until the infant had finished her meal – she just got up and went, and Pom, clinging for once under her mother, struggled to keep the nipple in her mouth as long as she could before she was relentlessly pushed up on to Passion's back. As a result of her mother's lack of solicitude, Pom seldom managed to suckle for more than two minutes at a time before she was interrupted by Passion, and often it was much less. Most infants, during their first years, suckle for about three minutes once an hour: Pom probably made up for her shorter feeds by suckling more frequently.

It was the same story when Pom started to walk. Flo, it will be remembered, was very solicitous when Flint was finding his feet, gathering him up if he fell and often supporting him with one hand as he wobbled along. Melissa was less concerned and, if Goblin fell and cried, merely reached her hand towards him whilst he struggled to his feet. But Passion was positively callous. One day, when Pom had never been seen to totter on her own for more than two yards, Passion suddenly got up and walked away from her infant. Pom, struggling to follow and falling repeatedly, whimpered louder and louder until finally her mother returned and shoved the infant on to her back. This happened again and again. As Pom learnt to walk better, Passion did not even bother to return when the infant cried – she just waited for her to catch up by herself.

When Pom was a year old it was a common sight to see Passion walking along followed by a whimpering infant who was frantically trying to catch up and climb aboard her moving transport. It was not really surprising that, during her second year, when most infants wander about happily quite far away from their mothers, Pom usually sat or played extremely close to Passion. Indeed, for months on end she actually held tightly on to Passion with one hand during her games with Flint and Goblin and the other infants. Obviously she was always terrified of being left behind.

Like human children, chimpanzee children are dependent on their mothers for several years. Most chimp youngsters continue to suckle and sleep with their mothers for over four years and, though they ride less and less often on their mothers, they are quick to jump aboard at any sign of excitement in the group or some other danger until they are four or even five years old. During this period of dependence the infant gradually masters his physical environment: he learns to move easily and rapidly over the ground and through the trees, and he becomes increasingly skilful in his manipulation of objects – such as branches and twigs whilst he is feeding and nest making.

Flint first attempted to make a nest when he was ten months old.

He bent a little twig over and sat on it on the ground in the approved manner. Then he bent a handful of grass stems *on to* his lap.
After this I sometimes saw him trying to make nests as he dangled
in mid-air, bending down twigs and attempting to hold them under
him with his feet as he reached for more. During the next few
months Flint became more and more proficient and, like other
year-old infants, he often made a nest whilst he was playing around
by himself in a tree. Sometimes he lay in it for a short while, but
more often he just bounced around in it, often breaking it apart
and then, after a few minutes, making another. This constant
practising means that when a youngster is four or five years old
and ready to sleep on his own he is skilled in nest-making techniques. It is the same with the use of twigs and sticks for insect-
eating: infants play with and poke about with such materials long
before they are interested in using them for the serious business of
feeding.

Human infants, by and large, learn to toddle and climb the stairs
and feed themselves with a spoon much more readily than they
learn to behave in the well-mannered way required of them by
most grown-ups. Quite often a small child shows a remarkable
lack of perception of the mood of his elders; when he has been
banging the table with a tin plate for minutes on end, his mother
may have to tell him repeatedly to stop, or snatch the plate away,
or smack him before he recognises her irritation. When his father
is engrossed in a book, the child may persistently try to attract his
attention, despite the fact that his only rewards are stern glances
and angry demands for him to keep quiet.

Such behaviour is sometimes labelled deliberate naughtiness: yet
we see the same sort of thing occurring in chimpanzee society.
Chimpanzee infants can walk and climb before they start to use
most of the submissive gestures of the adult communicatory
system during interactions with their elders, and they too, during
the first year of life, show a surprising lack of perception as to the
mood of their elders. At this time it is the chimpanzee mother who
must keep a watchful eye not only on her child but also on the
other individuals around them.

One day I followed a small group up the mountain-side. Presently
they settled down to groom and rest, and I sat down near them.
Goblin, still tottery on his legs at ten months, wobbled over to

Mike who was sitting in the shade of a palm tree chewing on a wadge of fig seeds and skin. As Goblin looked up at him the dominant male reached out and patted the infant's back very gently, time and time again. Goblin stumbled away, tripped over a length of vine, and fell flat on his face. Almost at once Fifi hurried over and gathered him up, holding him closely for a moment. She began to groom him but he pulled away and wandered off again. A small tree stump lay in his way and, fixing this with the intense concentrated gaze of the young infant, Goblin went up to it and tried to climb on to the top. Half-way up he lost his grip with one hand and nearly fell, but David Greybeard, who was sitting nearby, quickly reached out and put his hand under the infant's elbow until he was safely up.

Just then Flint, six months older than Goblin, came bouncing up and the two children began to play, both showing their lower teeth in the chimpanzee's playful smile. Flo was reclining nearby grooming Figan; Goblin's mother, Melissa, was a little farther away, also grooming. It was so peaceful, this scene in the deep forest, high up on the side of a mountain with the gleam of the lake just visible through a gap in the trees. But all at once a series of pant-hoots announced the arrival of more chimpanzees, and there was instant commotion in the group. Flint pulled away from the game and hurried to jump on to Flo's back as she moved, for safety, half-way up a palm tree. I saw Mike with his hair on end beginning to hoot; I knew he was about to display. So did the other chimpanzees of his group – all were alert, prepared to dash out of the way or to join in the displaying. All, that is, save Goblin. He seemed totally unconcerned and, incredibly, began actually to totter towards Mike. Melissa, squeaking with fear, was hurrying after her son, but she was too late; Mike began his charge and, as he passed Goblin, seized him up as though he was a branch and dragged him along the ground.

And then the normally fearful, cautious Melissa, frantic for her child, hurled herself at Mike. It was unprecedented behaviour, and she got severely beaten up for her interference. But she did succeed in rescuing Goblin; the infant lay, pressed close to the ground and screaming, where the dominant male had dropped him.

Even before Mike had ceased his attack on Melissa, the old male Huxley had seized Goblin from the ground. I felt sure he too was going to display with the infant, but he remained quite still, holding the child and staring down at him almost, it seemed, in bewilderment. Then as Melissa, screaming and bleeding, escaped from Mike, Huxley set the infant on the ground. As his mother hurried up to him Goblin leapt into her arms and she rushed off into the undergrowth: their screams gradually receded as Melissa hurried away from the group.

It is difficult to understand Mike's behaviour. Normally small infants are shown almost unlimited tolerance from all other members of the community, but it seems as though the adult male, during his charging display, may lose many of his social inhibitions. Possibly Mike, in the frenzy of the moment, considered any object lying in his path was a suitable prop to enhance his display. Once I watched Rodolf pound and drag an old female during one of his displays whilst her infant clung screaming beneath her; then, almost before he had stopped attacking her, he turned round to embrace, pat and kiss her.

It was only a couple of weeks after Mike had displayed with him that Goblin was involved in another dramatic sequence of events. It started whilst Goblin was playing with a female infant quite close to their two mothers who were grooming each other. Suddenly there was a commotion; a male charged towards the group, and one of the mothers was briefly attacked. Instantly both females raced towards their infants and Melissa, who got there first, seized the wrong one and rushed with her up the slope. The second mother grabbed Goblin but almost at once pushed him from her and hurried after her own screaming child. Goblin was left alone, his face almost split in two by his huge grin of fear. Just then Mike ran up, but this time his behaviour was completely different. He gathered up the terrified Goblin in the gentlest manner and carried him for some way pressed to his breast. When the infant struggled to escape Mike put him down, but stayed with him for ten minutes, threatening or chasing off other chimpanzees who approached too closely. When Melissa herself finally arrived on the scene, Mike watched benignly as Goblin raced to join his mother.

Goblin, like some other first infants of young and inexperienced mothers, got into more scrapes than a youngster such as Flint who had a mother to whom child-raising was no novel experience. Of course, when Flint was small there was Fifi to look after him too;

That's one way of looking at it

Flint and Goblin interfere when Faben mates

Olly, followed by Gilka, responds to Leakey's courtship

Fifi presses her bottom to the bottom of a mating male (See p. 167)

Adolescent males are usually nervous about joining in adult grooming sessions (See p. 163)
An experienced mother allows other chimpanzees to inspect her newborn infant (See p. 139)

several times it was Fifi who was closest when danger threatened, and she who seized him up and ran with him to safety.

Some mothers appear to be over-cautious, and repeatedly 'rescue' their infants from situations which do not seem to be dangerous at all. I remember, when Gilka was just a two-year-old, how excited she always got on the rare occasions when her mother remained, for a while, in a big group where there were a number of adult males. Like a little human girl showing off in front of a circle of grown-ups, Gilka would stand upright, swinging her arms and stamping her feet, or pirouetting round and round. If she approached one of the males he usually responded by reaching out to pat her or to tickle her in the most tolerant way possible; yet Olly nearly always hurried up, pant-grunting nervously, touched the male submissively, and took Gilka away. Once, when Gilka kept pirouetting over to a peaceful group of adult males, Olly four times followed and, quite literally, dragged her daughter away by the hand.

It was even worse, so far as Olly was concerned, when Gilka tried to initiate play with a mature male. Most males responded readily enough to the advances of the gay infant but, as soon as Olly noticed, she hurried up and either took Gilka away or began to groom the male so that, quite often, his attention was distracted from her child. Yet on none of these occasions were any of the males showing signs of aggressive behaviour.

Hugo and I will always remember the time when Olly, suddenly noticing that Gilka was having a tickling game with Rodolf, quickly approached and, with worried pant-grunts, laid her hand placatingly on the big male's back. Rodolf, who did not need placating, continued to play, lying on his side and heaving with quiet chimpanzee laughter as Gilka clambered up his large frame and play-bit into his ticklish neck.

After watching the two for a moment, Olly started to groom Rodolf, nervously pulling away her hand every time he waved an arm or leg during the game. Quite suddenly, Rodolf turned to Olly and began to tickle her in the neck with one hand whilst, with his foot, he continued to tickle Gilka. Olly's face was a study; her lips wobbled in agitation, her eyes stared, and her pant-grunts became positively hysterical. She pulled back, but Rodolf followed and tickled her again. For a few moments, her long lips bunched themselves into the semblance of a play face, and I detected a note of laughter in her frenzied grunts. But after a moment she could endure it no longer and moved away from such alarming physical contact with an adult male.

Sometimes an adult male does suddenly attack a female during a game. I am not quite sure why, but I think it is because, as he gets rougher and perhaps hurts the female, she draws away from him; this may irritate them in some way. But Olly, I am sure, need not have worried; I never saw Rodolf end any play session aggressively.

Young chimpanzees spend a great deal of time playing. Indeed, when they are two or three years old it often seems that they do little else. Play is a much argued about category of behaviour in scientific circles. What is it? What is its function? How should it be defined? And yet, despite all the discussions and speculations, most people, whether they be trained scientists or casual observers, readily agree as to when animals are playing, be they watching a litter of puppies tumbling about or a young chimpanzee turning somersaults.

Perhaps some of the arguments have arisen because in *human* children two entirely different types of activity are labelled 'play.' The two-year-old who, with intense concentration, builds ten blocks into a tower, is said to be playing with his bricks. It is a completely different sort of behaviour to that shown by the same child when he toddles round the sofa, shrieking with laughter, whilst his father crawls after him, patting at his legs. The infant chimpanzee who tries, again and again, to bend a branch under him for a nest or who attempts to catch a termite with a minute and totally inadequate piece of grass, is probably performing a behaviour that is equivalent to tower-building in the human child. But most of the behaviour which, in chimpanzees, we refer to as playful, is of the romping, laughing type shown by the human child when he is chased or tickled.

Young chimps often play by themselves when no playmates are available, swinging about in a tree, jumping over the same gap on to the same springy branch time after time, somersaulting or gambolling on the ground. Mostly, however, they like to play with each other, chasing round and round a tree-trunk; leaping, one after the other, through the tree-tops; dangling, each from one hand, whilst they spar and hit at each other; playfully biting or hitting or tickling each other as they wrestle on the ground.

Whether or not scientists ever agree as to the function of play, it certainly does serve, for one thing, to make the growing youngster familiar with his environment. He learns, during play, which type of branch is safe to jump on to and which will break, and he practises gymnastic skills, such as leaping down from one branch and catching another far below which, when he is older, will serve him in good stead – during an aggressive encounter with a higher-ranking individual in the tree-tops, for instance. It is not true, as some people have suggested, that he would learn these facts just as well during feeding, normal locomotion and so on, for these routine activities seldom necessitate wild leaps.

Over and above these considerations, social play certainly offers the young chimp the opportunity to become familiar with other youngsters. He learns which of them are physically stronger than he is; which ones have mothers higher-ranking than his own – who may retaliate, if a squabble breaks out, with unpleasant consequences to himself. He discovers which of his playmates can be intimidated by a show of strength and which of them will, in a similar context, turn round and call his bluff. In other words, he learns something of the complex structure of chimpanzee society.

However, whilst play may be a type of schooling for the young chimp, it is, quite obviously, a most enjoyable one. Many mothers have great difficulty in persuading their offspring to leave a game when they themselves are ready to move on. Passion, of course, seldom had any difficulty in persuading Pom to follow her – at least until the infant was three years old and less terrified of being left behind. At one time, indeed, Pom was so anxious lest Passion should go without her that she would sometimes rush away from her playmates if her mother merely moved a few yards to a new resting-place. Other infants, however, often showed great reluctance to stop a play session.

Flo sometimes coped with the task of wresting Flint from his playmates by playing with him herself. Then, when she dragged him away by one foot, he apparently continued to regard it as a game, for he often laughed as his back bump, bump, bumped over the uneven ground. I was always reminded of Christopher Robin dragging Pooh bear downstairs. But it was Melissa who really amused us. One day, when Goblin was playing with Flint, Pom and a couple of others, Melissa wanted to leave. She pulled Goblin from the group, pressed him to her tummy, and set off. She had gone only ten yards or so when Goblin detached himself and came cantering back to dive into the wrestling youngsters once more. Melissa, uttering soft whimpers, followed him. For a few moments she stood looking at her son and then, once more, she gathered him up and set off. This time she got farther, but when she had gone thirty yards Goblin again came romping back. Melissa followed, whimpering a little louder. She was still whimpering when she pulled him from the group and set off for the third time. Melissa took Goblin away, and followed him back, whimpering, no less than fifteen times before she finally succeeded in leaving.

When one watches the gradual development of male and female infants there are not many obvious differences in the behaviour of the two sexes. Male infants tend to indulge in more rough and tumble play than do females, and they practise aggressive display patterns more often during their games – such as dragging branches and swaggering about. Also male infants usually start to threaten and attack others at an earlier age than do females. There is, in addition, one major difference, and that lies in the precocious sexual development of the male infant.

From a very tender age the male shows great interest in the pink swellings of females. I remember Flint struggling to reach a female in this state almost before he could walk. Once he did get to her, he made repeated attempts to mount her as she reclined on the ground. At that time I was astonished, but subsequently it became clear that this was normal behaviour for a male infant, although, to be sure, Flint was somewhat forward.

Between the ages of one and four or five years, males tend to spend a great deal of time, when a pink female is in the group, hurrying up to her, mounting her, and making all the movements shown by an adult male during mating. I remember Goblin, when he was two years old, approaching one female fifteen times in half an hour. Usually she crouched for him so that he could mount her easily – a couple of times she stood up before he had finished with her so that he was suddenly lifted from the ground and ended by scrambling on to her back where he sat, looking rather surprised. When his approach did not cause the female to crouch down,

Goblin tried to reach her by climbing little twigs or by balancing on the tips of his toes as he 'mated' her.

Flint, when he was three years old, began to show some of the courtship displays of the adult male. One day, when he had 'mated' young Pooch about twenty times during a play session, she appeared to get tired of the infant and climbed away from him into a little tree. Flint sat below, staring up at her with all his hair on end, and waved twigs violently until, whimpering, she climbed down and once more crouched low for her small but demanding suitor.

When it comes to real sex, male and female infants behave in exactly the same way. Nearly always, as soon as they notice a mature male mating a female, they rush up and try to interfere. At no time does the male demonstrate more clearly his amazing tolerance towards infants. Sometimes he may be almost lost to view as up to four youngsters converge upon him, pushing at him or putting their hands to his face as he mates. For the most part he simply turns his head away and appears to try to ignore their very existence. Only occasionally does he hit out at an infant – and then the victim is usually a male of three or four years old.

Why infants should behave thus we do not know. When Fifi pushed at Flo's suitors during her infancy it seemed that something akin to jealousy might be motivating her, but since then we have seen males pushed and mobbed by infants whether the female happened to be the mother of one of them or not. For the time being it must remain one of the many mysteries.

During the chimpanzee youngster's fourth year the very tolerant atmosphere in which, up to this time, he has been nurtured, gradually begins to change. His play sessions become rougher and wilder, and older chimpanzees are quicker to threaten him if he behaves incautiously. This is the time, too, when most youngsters are actually weaned, and weaning can be a very trying business indeed, lasting, in some cases, for over a year. For Gilka, whose weaning period coincided with so many other adverse circumstances, the transition from infancy to the juvenile stage was a particularly unhappy time.

## Chapter 13

The Child

Listlessly Gilka dangled from the branch of a tree above me, one leg bent at the knee and the foot pressed into the opposite groin. For a full minute she remained thus, almost motionless. Then, her movements slow, she climbed to the ground and hobbled forward on three limbs, still keeping her foot tucked into the opposite groin. When she was some four feet from her mother, Olly, who was working with a grass tool fishing for termites, Gilka stopped and began to utter a series of low whimpers. For a minute Olly ignored the child and then she pulled Gilka towards her and began to tickle her. Soon Gilka was uttering the panting chuckles of chimpanzee laughter. But the game lasted for less than a minute – then Olly pushed her daughter away and resumed her endless termite fishing. Nevertheless, just as Fifi had been temporarily distracted from her early attempts to touch or pull away Flint by similar playful behaviour on Flo's part, so it was with Gilka. She looked around, picked up one of Olly's discarded tools, and idly pushed it into an opened-up termite passage. No insect clung to the end when she withdrew the grass; she tried once more and then abandoned the attempt and sat grooming herself.

A few minutes later Gilka again approached Olly and stood, uttering her soft pleading whimpers. For a while Olly ignored her completely, then suddenly she reached out and drew her daughter close, allowing her to suckle for about half a minute before pushing her away. Gilka stood staring at Olly for a moment, then turned away and climbed back into her tree. There she sat, slumped, and picked little pieces of bark from the trunk. She broke them in her fingers, scarcely looking at what she was doing, and dropped them to the ground.

At that time Gilka was about four and a half years old. For the previous seven months she had been going through an increasingly difficult period. First of all her elder brother Evered, who had been her constant playmate, had reached adolescence and, in consequence, moved about with his family less and less frequently. Secondly, Fifi's attitude towards Gilka had undergone a sudden change. When Flint was three months old Fifi had not only ignored most of Gilka's invitations to play, but also threatened and sometimes attacked the younger chimp whenever she went too close to Flint. And finally, just as Fifi became less possessive of her small brother, more willing to play with Gilka, more tolerant of Gilka playing with Flint, Olly began to shun the Flo family. This was almost certainly due to the fact that, at the time, Faben and Figan were both spending many days with Flo and the rest of the family. Olly was nervous when in the company of these two physically splendid and vigorous young males.

So Gilka, often for days at a stretch, wandered through the forests with only her old mother for companionship – a mother who was weaning her child and rejecting Gilka's attempts to suckle with increasing determination. Also it was the termiting season, and Gilka, like all youngsters, got bored when her mother sat fishing for the insects for three or more hours at a time.

Small wonder that this combination of circumstances resulted in Gilka, formerly such a gay and lively little chimp, becoming increasingly lethargic. Small wonder that she began to show strange idiosyncrasies – such as the putting of one foot in the opposite groin, sometimes for minutes on end, or senselessly doodling with pieces of bark. Almost certainly it was because of her boredom, her lack of chimpanzee playmates, that Gilka, during that period, formed a very strange friendship indeed.

One day, when Gilka was again hanging around whilst Olly fished for termites, I heard a baboon bark farther down the valley. At the sound Gilka's whole attitude underwent a sudden change. She stood upright, peering towards the sound, then climbed higher in the tree and peered towards a clearing about one hundred yards down the valley. When I too looked in that direction, I could just make out some baboons moving through the trees. After a few moments Gilka swung rapidly from the tree and set off towards the clearing. Olly merely glanced after her daughter and then went on with her endless termite eating.

I followed Gilka for a short way and then, when I could see

fairly well into the clearing, stopped to watch what would happen. A moment later I saw Gilka move out from the trees; and at almost the same time a small baboon detached itself from the troop and cantered towards her. I did not need binoculars to identify Goblina, a female of about the same age as Gilka. As I watched, the two ran up to each other and, for a few seconds their faces were very close together. Each had one arm around the other. The next moment they were playing, wrestling and patting at each other. Goblina went around behind Gilka and, reaching forward, seemed to tickle the chimpanzee in the ribs. Gilka, leaning back, pushed at Goblina's hands, her mouth open in a wide smile.

It is fairly common for young chimpanzees and young baboons to play together, but the games usually consist of wild chasing around, either on the ground or through the trees, or sparring when each hits out quickly towards the other and then draws away. Often, too, such games end with aggressive behaviour from one or the other. Gilka's friendship with Goblina was quite different: the contact between the two youngsters was nearly always gentle, and they often deliberately sought each other's company – as they had done on this occasion. No one was studying baboons at that time, but Hugo and I had known Goblina for over a year and it seemed that, at quite a young age, she had lost her mother. When I watched her troop settle down for sleep one night I saw Goblina run from one adult female to another, and finally huddle close to an old and childless female. But unlike Gilka, of course, she had baboon playmates galore in her troop.

I watched Gilka and Goblina playing for ten minutes, and all the time they were amazingly gentle. Then the baboon troop started to move on and Goblina scampered after it. Gilka stood looking after her and then turned and slowly walked back towards Olly. She swung into a tree as she passed me and, still playful, stamped along a branch just above my head, showering me with twigs. Then she approached Olly and began her soft whimpers. This time, however, Olly continued to ignore her daughter and, after a while, Gilka climbed into a tree. The playful mood had gone and she began picking off pieces of bark, crumbling them in her fingers, and dropping them to the ground.

The strange friendship between Gilka and Goblina lasted for nearly a year; then Olly and Gilka disappeared. For a while we thought something must have happened to them, but then they were seen some three miles to the north. When they returned to our valley, six months later, Goblina was already adolescent and less playful, for baboons mature faster than chimps. The friendship between the two youngsters was not revived.

We soon realised that during those six months Gilka had been weaned, and although she still accompanied her mother constantly, we saw few friendly interactions between the two. Indeed, Olly, whom we soon discovered was pregnant, sometimes seemed to be unnecessarily aggressive towards her daughter. Often, for instance, she would mildly threaten Gilka if the child approached within ten feet or so when she was feeding – even when the two were eating in a tree that was bearing more than sufficient for their needs.

I was interested to find out whether Gilka was still sharing her mother's nest at night and so, at the first opportunity, I followed the pair when they left the feeding area after a late afternoon visit. In the past I had spent many hours roaming the forests with these two chimpanzees, and they paid me scant attention as we walked briskly along one of the well-worn tracks leading into the mountains. Occasionally Olly or Gilka paused to pluck a ripe fruit or a tempting handful of vine leaves from the side of the path, but obviously they had a definite goal in view.

Presently we left the forest and moved on to one of the ridges overlooking the lake. There the grass was high – on a level with my head. Often I feared I had lost Olly and Gilka, but luckily the faint rustle they made as they proceeded along gave away their whereabouts, and so I managed to keep behind them. Just before dusk, Olly, closely followed by Gilka, climbed into a tall tree. For twenty minutes the two of them fed on the yellow blossoms that grew in profusion. I found a comfortable rock, still warm from the sun, and settled down to wait until their meal was over. I had a view over the evening lake and watched as the reflected crimson and brick-red of the sunset gradually gave place to bluish-purple and steel-grey as the sun sank lower and lower behind the dark mountains of the Congo on the far side of the lake. The high-pitched shrilling of the cicadas gave way to the night symphony of the crickets. Slowly the colours drained from the trees and grass and the thin sickle of the new moon and her attendant evening star became visible above the lake. Would Olly and Gilka never finish their evening meal?

When there was just enough light for me to see the tree ahead, Olly and Gilka suddenly climbed down and set off along a narrow track towards a small pocket of forest some one hundred yards distant. Hastily I followed; when they entered the trees the blackness of their coats merged with the darkness and I could no longer see them. I knew, however, that it would only be a matter of moments before they went to bed, so I walked a short distance along the track and then stopped to listen. Sure enough, I soon heard the loud report of a snapping branch and, moving my position a little, saw the dark shape of a chimp outlined against a sky from which the last vestiges of daylight were rapidly fading. Two minutes later the nest was completed and the chimp lay down.

A few moments afterwards I saw movement in another part of the same tree as a smaller chimp began to construct a nest. Presently Gilka also lay still. I waited another ten minutes, for it is not uncommon for an infant to construct a small nest near that of its mother in the evening and then leave it to join her. Gilka, however, did not move, so I set off for camp, grateful, in the darkness, for the torch which is always a part of my equipment. I have always found walking through tall grass rather unnerving at night. It is actually easier to find one's way about without a torch; even when there is no moon the light from the stars is usually sufficient to outline the salient features of the landscape, and these are no longer clear once a light is switched on. In my early days at Gombe Stream, however, I had learned that the small circle of light provided by a torch gave me a tremendous feeling of security. Beyond in the outer darkness, the leopard could prowl and the buffalo stamp and all to no avail, for, within my own magic light spot, where the grass regained its colour and the rocks on the ground their shape, I was safe. What a difference the discovery of fire must have made to the life of primitive man.

I followed Olly and Gilka one other evening, and again Gilka settled down for the night in her own separate nest. That time I returned before dawn and saw that Gilka was still in the same nest when she woke up in the morning.

I described, earlier, how Fifi went through a period of dependence just after Flo's milk dried up and, for a while, clung to her mother like an infant. Her problems, however, had been of short duration compared to Gilka's and, once she had lost her almost fanatical obsession for Flint, she became a relaxed and particularly playful juvenile. She not only enjoyed games with all the members of her own family, but also with many of the adolescent and mature males. Once she romped with old J.B. for twenty minutes, chasing him round and round a palm tree. The large, fat and often bad-tempered male was soon laughing out loud as he ran, seemingly in expectation of the tickling that would come when he stopped and Fifi flung herself on to him. Most other juveniles that we have known, both male and female, were too scared of the big males to play with them except on rare occasions.

Possibly Fifi's relaxed behaviour with her elders stemmed from the fact that she enjoyed a particularly friendly relationship with her mother. Flo was more tolerant of her juvenile daughter than Olly was of hers, and far more so than old Marina of her juvenile daughter Miff. Miff was a contemporary of Fifi's; she had one elder brother and one younger. Their mother appeared to have a very cold disposition indeed. I never saw her playing with Miff; indeed, she seldom played with two-year-old Merlin unless he pestered her to tickle him, again and again pulling her hand towards him.

I never saw any friendly interaction between Miff and her mother except for their sessions of social grooming. Miff, in fact, was scared of her mother. She never ran up to share a box of bananas with Marina, as Fifi always did with Flo. Nor did Miff beg for fruit from her mother if she herself had none, whereas Fifi not only begged persistently in such a situation, but was liable to fly into a tantrum if Flo withheld the food – screaming, hurling herself to the ground, flailing her arms. This often resulted in Flo relenting and handing her daughter a banana – at least until Fifi was an adolescent of about eight years old.

I have particularly vivid recollections of the behaviour of Marina and Flo towards their respective daughters during the termiting season. Once, when Marina and Miff were working at the same heap, I noticed that Marina was having little success at her hole whilst, a few yards away, Miff's hole was working well. Suddenly Marina walked over to her daughter and deliberately pushed her aside. Miff moved, whimpering slightly, and watched as Marina extracted a grass-full of juicy termites. Then the child moved some yards from the heap, selected another long grass stem, and went to try her luck at a new hole. Just then Marina's own grass tool bent in the middle and, without hesitation, she reached over and pulled the fine new stem from Miff's hand.

What a contrast was Flo's treatment of Fifi. Once, near the beginning of the season, Flo chose to work a termite heap which was partially covered, with dead leaves. The openings of the nest passage were buried, and it took a while before Flo, scratching around, found a good place to work. Fifi wandered around looking too, but she could not find a suitable hole and finally went and sat close to Flo, peering intently. After a few moments Fifi began to whimper, rocking slightly back and forth, and inching her hand, in which she held a grass stem, towards her mother's hole. Presently, as Flo withdrew a stem, Fifi, looking up at her mother's face for a moment, cautiously poked her own tool into the hole. Patiently Flo waited until her daughter had pulled the stem out before once more poking her own down. This continued for a while, and then Flo moved off and found another place to work. A short while later, when Fifi's hole stopped yielding a good supply of insects, she again approached her mother and tried to share her hole. This time Flo, twice, gently pushed the child's hand away, but when Fifi began to whimper and rock to and fro, her mother gave in and once more went off in search of a new place to work.

The young juvenile, particularly if her mother has a new infant, has to learn that it is now up to her to keep an eye on her mother and not, as in the past, the other way round. If she does become accidentally separated, she usually gets very upset. In fact, old Flo nearly always waited for Fifi even after Flint's birth, but once, when Fifi was about five and a half, she was so busy playing with an infant that she didn't notice Flo getting up to go. Flo, after looking round many times, gave up and wandered away with Flint.

As soon as Fifi realised that her mother was no longer in the group she became agitated. Whimpering softly to herself, she rushed up a tall tree and ran from one side to the other, staring across the valley in different directions. Her soft calls of distress became loud grating screams. All at once she swung down and, still crying, hurried along a track; she chose the direction exactly opposite to that which her mother had taken. I followed Fifi. Every so often she climbed a tree and stared round and then, with her hair on end, started along the track again, crying and whimpering.

Just before dusk she came upon Olly and Gilka, but though Gilka kept approaching Fifi, grooming her, trying to play, Fifi ignored her friendly advances and, instead of remaining to sleep near Olly, hurried off by herself again. She made a lonely nest at the top of a tall tree. I stayed out close by, and three times during the moonless night I heard her calling out, screaming and whimpering.

Very early the next morning, when it was still dark, Fifi left her nest and ran off into the forest, whimpering. It was not light enough for me to follow so I returned to camp. Hugo told me that Fifi had appeared, still whimpering and staring around, at about seven o'clock that morning, and had then hurried off up the valley.

Two hours later Fifi arrived back in camp with Faben. Normally Faben paid little attention to his sister, except to join her occasionally for a game, but Fifi now seemed calm and relaxed in his company. They were still in camp when Flo arrived. We expected a rapturous reunion, but Fifi merely hurried over to her mother and the two settled down to groom each other, intently and vigorously. Since then we have seen other reunions between mothers and their lost children; each time the greeting has been unspectacular, but the two have groomed each other – another indication of the importance of social grooming in chimpanzee society.

One thing which puzzles us is that whilst a mother will hurry towards the sound if she hears her lost child crying, she herself makes no loud call which would indicate her whereabouts to the child. And so if, when she reaches the source of the crying, the child has moved out of earshot, it may be hours, or even days, before the two are reunited. Marina, as might be expected, seldom looked to see whether Miff was following when she left, and Miff's sad whimpering and crying as she searched for her mother were very common sounds indeed when she was a five-year-old.

The male juvenile, whilst he also gets upset when he accidentally loses his mother, nevertheless may initiate his independence at a much earlier age than a female. Some juvenile males may move about with other chimps for several days at a time when they are only six years old.

Figan, however, spent most of his childhood roaming the mountains with Flo and Fifi. One day, when he was about six years old, he was with Flo and Fifi feeding in a large fig tree. Suddenly a loud burst of pant-hooting and drumming announced the arrival of a large chimpanzee group in the valley. From the sounds it was obvious that they were climbing into some more fig trees farther up the stream. Flo and her family looked towards the sounds and hooted, and Figan swung through his tree, stamping on the branches, and so to the ground. He ran along the track towards the others, again breaking into pant-hoots and then drumming on the buttress of a tree. After this he paused and looked back towards Flo, obviously expecting that she would follow. But the old female was quite content where she was.

Presently Figan turned and went on but, where the track wound out of sight, he again stopped and glanced towards Flo. After a moment he began to walk back towards his mother; then he stopped, looked once more towards the big group, turned round and set off resolutely away from Flo and Fifi. This time he vanished from sight and I heard his high-pitched pant-hoots farther along the track.

Within four minutes, however, he was back, walking in a non-chalant manner and hitting out at me, half playfully half aggressively, as he passed; the sort of gesture by which a human might try to cover a feeling of embarrassment.

Later on in the day Flo did join the big group. When she left, to make her nest farther away, Figan, quite deliberately, chose to stay behind. He was away from his mother for two days and they met again in camp. Fifi, who had missed her playmate, rushed up and embraced him, but Figan ignored her. Then, to complete the picture of the undemonstrative elder brother, he walked over to Flo and merely brushed the side of her face with his lips. Soon, though, he was tumbling about with his sister in evident enjoyment.

Male juveniles show much caution when interacting with their elders. It is probably their increasing respect for the adult males that leads to a reduction of, or complete abstinence from the mounting of pink females that occurs so frequently in male infants. Whilst I have never seen a juvenile son who was frightened of his mother, as Miff was of Marina, nevertheless, the male juvenile normally continues to show a good deal of respect for his mother.

One day I came across Figan with the freshly killed body of a colobus monkey. He climbed a tree with the tail of the monkey in one hand and the body slung over his shoulder. He was closely pursued by Fifi. When he reached a comfortable branch he sat and began eating, and Fifi, who was about three years old at that time, begged persistently. Several times Figan gave her small fragments of meat.

A few minutes later I saw Flo climbing towards Figan. Instantly he slung the carcass over his shoulder and climbed to a higher branch. Flo remained in a low fork, gazing round, not once looking at her son. He relaxed and began feeding again, though constantly he glanced, somewhat apprehensively, towards his mother. The old female sat there for a full ten minutes. Then, with only a preliminary fleeting glance at Figan, she very slowly climbed a little higher, looking oh so nonchalant, and sat on the next branch up. Figan, looking down even more frequently, continued to feed. But the next time Flo casually climbed a little higher, Figan, feeling perhaps that she was getting too close, climbed higher too.

And so it went on. Flo's intentions were obvious, to her son as well as to me. When he reached a very high position in the tree, Flo could keep up her pretence no longer – suddenly she rushed towards him and Figan, with a scream, leapt down into the foliage and vanished from sight, with Fifi and Flo swinging in pursuit. I could not find them again.

# Chapter 14  The Adolescent

Adolescence is a difficult and frustrating time for some chimpanzees just as it is for some humans. Possibly it is worse for males – in both species. The male chimpanzee becomes physically mature at puberty, when he is between seven and eight years old, but he is still nowhere near full grown – he weighs about forty pounds as compared to the hundred pounds of the fully mature male. And he is still far from socially mature – indeed, he will not be ranked amongst the mature males for another six or seven years. He is increasingly able to dominate, even terrorise, females – yet in his interactions with the big males he must become ever more cautious in order to avoid arousing their aggression.

One of the most stabilising factors for the adolescent male may well be his relationship with his mother. Old Flo, affectionate and tolerant with all her offspring, was frequently accompanied by Faben and Figan during their adolescence. Olly and Marina, who were less relaxed and tolerant of their youngsters than was Flo, were joined less often by their adolescent sons but, nevertheless, we saw them together on many occasions. For the most part, these adolescent males, even when they were ten or eleven years old, continued to show respect for their mothers. If we offered a banana the son usually stood back and waited for his mother to take the fruit. Once I held a banana between Evered and Olly: both hesitated, and then both reached forward. Evered at once drew back his hand – but so did Olly, her lip wobbling as she turned quickly to glance at her son. Very gradually Evered's hair began to stand on end, but he made no further move and Olly, with a series of nervous pant-grunts, finally reached for the banana.

On many occasions a mother will hurry to try to help her adolescent son. Once when Mr Worzle attacked Faben, who was about twelve at the time, Flo, with Flint clinging to her and her

Faben brandishes a stick on seeing his reflection in a mirror

Faben throws a stone at a baboon who has approached a banana box

Figan tries to play with the old baboon, Job (See p. 193)

Mike, the dominant male, begs Leakey for a piece of meat (*Photo* GEZA TELEKI)

hair on end, rushed towards the scene of strife. As she approached, Faben's frightened screams instantly turned to angry *waa* barks and he began to display, standing upright and swaggering from foot to foot. Then mother and son, side by side, charged along the track towards old Mr Worzle, with Flo, uttering loud barks in her hoarse voice, stamping on the ground in her fury. Mr Worzle turned and fled.

When an adolescent is attacked by a high-ranking male then, of course, there is little the mother can do, but she usually hurries up to see what is going on, and may utter *waa* barks in the background. Even timid Olly once hung about barking whilst Mike attacked Evered and afterwards, when her son had run off screaming, she crept submissively up to Mike, pant-grunting hysterically, and laid her hand on his back – as though to propitiate him for whatever rash act on the part of her son had led to the fight.

Of course the relationship between mother and son does change somewhat as the young male grows older. When Figan rushed past Flo in a charging display, at the age of seven, she completely ignored him – though other adult females normally fled. A year later, however, Flo too rushed out of Figan's way when he careered towards her, dragging a branch and, with his hair on end, looking twice his normal size. Nevertheless, during that same year, we once saw Flo, during the excitement of banana feeding, pound aggressively on Figan's back with her fists so that he ran off, screaming.

As the adolescent male gets older he becomes increasingly likely to hurry to his mother's aid when *she* is threatened or attacked. One day the old mother Marina threatened Fifi. Fifi screamed, Flo ran to her daughter's assistance, and the two old females tumbled over in the dust. At this moment Marina's nine-year-old son Pepe, who had been feeding in a palm tree, noticed what was going on. He climbed rapidly from his tree and charged towards his mother. Flo, seeing him coming, turned and fled. Then Marina and Pepe chased Flo and Fifi right away and Flo, who had been able to make mincemeat out of Pepe only two years before, screamed until she choked with rage.

In his dealings with the higher-ranking males, the adolescent must be cautious for now, more than when he was a mere juvenile, an act of insubordination is likely to bring severe retribution. Once, when Leakey was enjoying a particularly large pile of bananas, we saw Pepe moving very cautiously towards him. Pepe, obviously, was half expecting some threatening gesture warning him to keep his distance, for each time the mature male made a sudden movement Pepe jumped. Gradually, though, he got closer and closer and finally sat, only a few feet away, with a huge grin of fear on his face. He reached hesitantly towards a banana, but his fear of Leakey caused him to withdraw his hand with sudden nervous squeaks. Again he reached out, and again caution got the better of him and he pulled back, screaming louder. Then Leakey leant towards Pepe and reassuringly touched the young male on his mouth and then in his groin. Pepe, however, still continued to make little fear-squeaks until Leakey once more reached towards him and patted the youngster's head and face gently and repeatedly. At last Pepe was calmed and gathered a small pile of bananas to take away and eat at a more comfortable distance from the mature male.

Many adult males, however, are less tolerant than Leakey. Indeed, on other occasions, in a different mood, Leakey might well have threatened rather than reassured Pepe, however submissive the adolescent's approach had been. And there are times when, as a youngster sits, at a distance, watching a high-ranking male stuffing himself with food, the tension appears to build up until the adolescent has to give vent to his frustration in a charging display. Off he goes, crashing through the undergrowth and dragging branches. But even this may provoke a reprimand from one of the big males – for it seems that adult chimpanzees are often irritated by a lot of noise and commotion from a youngster. The adolescent may well be chased and even attacked for his ill-advised display. Figan learnt this lesson quickly. Twice, during the same week, we saw Mike attack him after such displays. The following week, after rocking slightly in frustration as he watched Mike feeding, Figan, who had been unable to get any bananas for himself, suddenly got up and half-walked, half-ran away from the group. As he went he gave loud whimpering calls, like a child, until he neared a huge buttressed tree that grew some hundred yards along the main track. His whimpers became a series of high-pitched adolescent pant-hoots and he leapt up at the buttresses of the tree and drummed with his feet rhythmically *de-dum, de-dum, de-dum*. Then he came walking back along the path in that jaunty manner of his and sat down again, seemingly quite relaxed and calm. Subsequently, Figan went off on his own to display and drum on that tree on many occasions when he was frustrated in the presence of his elders.

However, even Figan, with all his skill in avoiding punishment, was often attacked as an adolescent; and others of his age seemed to be attacked more frequently. Why, then, does the adolescent male so frequently associate with mature males? Part of the answer probably lies in the fact that on most occasions aggressive incidents are, as it were, 'made up' very quickly – particularly those involving *young* adolescent males. Indeed, for many youngsters, the need for friendly physical contact from a male who has just threatened or attacked them appears to be compelling.

Once Evered was attacked really savagely by Goliath – simply because he got in the way during a charging display. At the end of the incident Evered was bleeding and had lost great handfuls of hair. Yet even so he followed Goliath when the big male walked away and, when Goliath sat, crept cautiously towards him. Evered was screaming loudly, and his fear of Goliath was such that he kept turning and starting to move away; yet his desire for a reassuring touch prevailed and eventually he got close enough to present his rump, still screaming and crouched flat to the ground. After a moment Goliath reached out and began to pat Evered's back again and again until, after fully one and a half minutes, Evered's screams became whimpers, and he finally quietened altogether. Only then did Goliath stop patting him.

Of course, away from the artificial conditions of the feeding area, the aggression of the high-ranking males towards their subordinates is much decreased, and the young adolescent male often appears more relaxed in the company of adult males out in the forest. Usually, however, he does not join in the long sessions of social grooming so beloved by his elders: he sits a few yards away from the mature males, grooming himself. When the group climbs into a tree to feed, the adolescent male often sits at a discreet distance from his superiors – sometimes, indeed, he feeds in a neighbouring tree. When the mature males start their displays, on arrival at some food source, the adolescent male is likely to keep out of the way until things have calmed down. Nevertheless, he is a part of the adult male group and, therefore, able to learn from his superiors – and young chimpanzees almost certainly do learn a good deal simply from watching others of their kind. The growing female can learn much that will benefit her in later life from her mother particularly when a new sibling is born, but since the young male has no 'father' as such – no male, that is, who is attached permanently to the family group – he must leave his mother and deliberately seek out adult male company.

When the frustrations of being with individuals so dominant to him become too great, the adolescent male either travels with his mother for a while or, often, by himself. Nearly all the adolescent males we have known have, as they got older, spent long periods – hours or even days – completely out of sight and often out of earshot of other chimpanzees. This aloneness, on some occasions anyway, is quite definitely deliberate. I once followed Figan when, as an eight-year-old, he set off with a large group of adult males and females. For an hour he remained with the others and then, when they climbed into a tree to feed, moved on by himself. Presently he too stopped to eat figs; twenty minutes later he swung to the ground and travelled on, still heading away from the group. The following day he visited the feeding area, still on his own, and he was alone when he left. That evening, however, he joined up with Flo and Fifi.

We have watched several splendid young males, aged between thirteen and fifteen, gradually leaving adolescence behind them as they entered the dominance hierarchy of the socially mature males. Faben and Pepe, who were about the same age, not only displayed frequently and vigorously at each other, competing for dominance, but also began to display at the lower-ranking mature males. It may well be that the charging display is particularly adaptive in this respect. I discussed, earlier, the possibility that, during such a display, the male chimpanzee might lose many of his social inhibitions. This could explain why a young male may actually charge towards a chimpanzee to whom, during less frenzied moments, he would show great respect. Even a high-ranking male may, on occasions, move out of the way of a vigorously displaying adolescent. If an adolescent is able to displace low-ranking adult males sufficiently frequently it will undoubtedly increase his self-confidence; and it may well be that the pattern of dominance will be permanently disrupted. Certainly it seems that the more spectacular his charging display the more likely it is that the youngster will break into the hierarchy of his elders. After a while both Faben and Pepe began to hold their ground and even retaliate when lower-ranking adults threatened them. They started to join in grooming sessions with Rodolf and Leakey and David Greybeard, the more tolerant of the high-ranking males. They charged into camp along with the other big males, and took their turn at courting attractive females during the frenzy of excitement after arriving at a food source instead of waiting until things were calm and their elders less worked up. Their period of apprenticeship was over and, from then on, their rise in the dominance ladder would depend on individual intelligence and determination rather than the slow maturing of their physical selves.

\* \* \*

The female chimpanzee also becomes adolescent at about seven years of age. She starts to show small and irregularly occurring swellings of her sex skin, although she will not menstruate or become sexually attractive to the mature males for another two years or so. She, too, must be cautious in her dealings with her social superiors – in her case, not only the adult males but also the adult females and adolescent males as well. And she may be threatened by the precocious juvenile offspring of high-ranking mothers.

When Fifi became an adolescent there was no discernible change in her relationship with her mother except that, gradually, Flo became less tolerant of her daughter sharing in her food. Fifi continued to follow Flo through the forests and to help look after the four-year-old Flint. When Fifi was threatened or attacked Flo usually hurried to help and Fifi would sometimes back up her mother. So far we have only been able to observe interactions between one other mother with her adolescent daughter. Like Flo and Fifi, these two usually travelled around together, but for the rest their relationship seemed very different; the daughter was tense and nervous of her mother in feeding situations, and we never saw either of them show concern if the other was threatened or attacked. There was, in fact, little of the relaxed and friendly comradeship which existed between Flo and Fifi although, in both cases, mother and daughter spent much time grooming each other.

During her adolescence a female is, if anything, even more fascinated by infants than she was as a juvenile. Not only does she frequently carry an infant short distances, play with and groom it, but she also shows concern for its welfare. Once, when Pooch was about eight years old, she carried a six-month-old infant up a tree and began to groom him. It so happened that the tip of a palm frond, swaying gently in the breeze, occasionally brushed past Pooch's shoulders as she sat there, and I noticed that the infant was watching this frond, as though fascinated. All at once he grabbed hold of it, wriggling off Pooch's lap – and, with his weight, the frond swung away back towards the trunk of the palm tree. For a moment Pooch stared after the infant as though she couldn't believe her eyes; then her face suddenly contorted in a huge grin of fear. Frantically she climbed from her tree and rushed up another from a branch of which she was just able to reach the frond as it swung like a pendulum back towards her. Then she seized the infant – who was not at all frightened but probably thoroughly enjoying the swing – and hugged him tight. The grin of fright only slowly left her face.

After about a year of adolescence the sexual swelling of the female gradually becomes larger and larger – though it does not yet equal the proportions of the fully adult swelling. Even at this stage the mature males show no interest, though male infants constantly mount young females when they are pink. Indeed, the female, at this stage, seems to welcome such attention from her tiny suitors and is usually quick to present, crouching very low, when they approach. One young female, indeed, actually pulled Flint away when he was mating another, only to crouch to the ground as she solicited his attention herself.

Finally, however, the day comes when a female develops a swelling which attracts the mature males – usually when she is about nine years old. I well remember the day when Pooch showed her first grown-up pinkness. First one and then another of the adult males approached her with his hair erect, shook branches at her, swaggered from foot to foot or hunched his shoulders – for almost all of the gestures of courtship are also part of threatening behaviour. Again and again Pooch screamed and fled, but the males followed her and their courtship became increasingly aggressive. Finally Pooch turned and rushed towards them, crouching close to the ground and screaming as, at last, the males mounted her. After each sexual act she rushed away, still screaming. Her fright and bewilderment that first day were only too obvious. By the second day, however, she had calmed down somewhat, and though she still screamed when a male approached her, she appeared far less terrified.

Fifi's sexual development was not only quite different to that of Pooch, but also different to that of any other young female we have watched to date. In fact, she was possibly the chimpanzee equivalent of a nymphomaniac. About six months before she developed her first truly adult swelling, Fifi showed an almost fanatical interest in the sexual behaviour of older females. Sometimes she would trail around after Pooch or Gigi, or one of the others when they were pink, so that she was close at hand when they were mounted. And, when such moments came, Fifi either jumped on to the pink female's back, pushing her own little bottom as close to the male as she could, or she rushed around behind the pair and pressed her bottom against the male's as he mated.

When Fifi finally achieved a full swelling of her own she showed none of Pooch's terror but responded instantly to the slightest sign of sexual interest in any of the males. Indeed, she often forestalled them, hurrying up and presenting before they had even looked in her direction. When these first pink days finally ended, Fifi, it seemed, could not believe it. The first morning we saw her hurry up to Mike, turn and present her rump, and crouch, soliciting mating. For a few moments she remained motionless, then she looked at him over her shoulder as though wondering why he did not respond. She backed an inch towards him, still watching. Finally Mike reached out, made a few grooming movements on her rump, then moved away. Slowly Fifi sat up, staring after him as though in amazement. Presently Evered came into camp. Fifi rushed over to him and repeated her solicitation. Again and again she backed towards him, again and again he inched out of the way. The last time she backed he had just turned to look elsewhere, and Fifi caught him off balance. Evered all but toppled over backwards before scrambling to his feet and making haste to increase the distance between himself and this importunate young female!

Fifi continued to present in this way for a couple more days and then, it seemed, resigned herself to the fact that she was no longer attractive. We couldn't help laughing when, during her next period of pinkness, we constantly saw her reclining on her side with one hand draped, as though protectively, over her pink swelling – for all the world as if she was determined that *this* time it would not vanish so mysteriously.

For the next year Fifi continued to show a keen desire for sexual contact during her pink days. She took to hanging about, either in camp or somewhere in the valley, where she could observe the chimpanzees who arrived for bananas. If a male arrived she would often rush over and solicit mating. Whilst the adult males did not show quite the enthusiasm for the daughter as they had for the mother, they usually responded to her invitations.

We were interested to discover that Fifi was extremely reluctant to be mated by her brothers. She even prevented little Flint from mounting her – though in the days before her first true swelling she had shown no objection whatsoever. Moreover, though Figan and Faben were observed to mate with their sister, after much screaming on her part, subsequent sexual interactions between the siblings only occurred very rarely.

There was one chaotic period when Flo and Fifi went pink together. The males had an exhausting eight days or so and the chimps moved about in a huge group of over twenty individuals. Every time that Flo was mated Fifi and Flint rushed over to interfere, pushing at the male's face and, afterwards, Fifi would solicit mating herself. And every time that Fifi was mated Flo and Flint would rush up and interfere, after which the male would usually mate with Flo. Added to this there were usually two or more infants other than Flint rushing up to interfere as well, so that sometimes the mating couple would be all but invisible beneath the cluster of chimpanzees that were all getting in the way. It was one of the rare occasions when chimpanzees other than infants were observed interfering with mating.

It was possibly significant that not once, during these days of intense sexual activity, did we see either Faben or Figan try to mate with their mother. And this despite the fact that they were a part of the group surrounding Flo and that every other physically mature male did copulate with Flo. Nor did we ever see Evered attempt to mate with his mother when Olly was pink.

Of the nine females whose progress during adolescence we have been able to watch, not one has given birth until at least two years after her first adult swelling. The same is often true of captive individuals, but exactly what physiological processes are at work we do not, as yet, understand. Certainly, though, the time lag is beneficial to a wild chimpanzee female, for at nine years of age she is not socially mature, nor is she completely fully grown, and she has quite enough to cope with without the added burden and responsibility of a baby.

# Chapter 15 Adult Relationships

In human societies there is no definite point in time at which the pimples and awkwardness of youth gives place to the assurance of young woman- or manhood. The change gradually takes place over a period of many months; one day, to their surprise, the parents of a child suddenly realise that he is a child no longer. It is the same with the chimpanzee.

One hot day in the summer of 1966, as I watched the old mother Marina and her children, I suddenly realised that Pepe was, indeed, a splendidly mature male; an adolescent no longer.

I could see the muscles taut beneath his sleek coat as, with both hands, he pushed a short thick length of stick backwards and forwards in the entrance of an underground bee's nest, enlarging the opening. Clustered around him were two females; his mother, Marina, and his sibling, the juvenile female Miff. All three seemed oblivious to the swarm of infuriated bees buzzing around them, but the infant of the family, little Merlin, had rushed up a tree some distance away when the others had began raiding the nest for honey and grubs.

After a few moments Pepe put his stick aside and waited while Marina reached down with her hand and withdrew a section of honeycomb. Then he reached in and took out some crushed waxen cells for himself. Miff did not dare take any of the treat, but she watched intently as the other two ate, her face only an inch or two from Pepe's mouth.

Fifteen minutes later Pepe wandered away from the empty nest. Marina felt inside once more before following him, and Miff, finally, got her hand down into the hole and time and again licked from her fingers the slightly sweet earth. Then she too followed her brother into the shade of a large tree. Merlin swung from his branch and made a large detour, moving through the trees, to join the rest of his family.

Pepe began to groom with Marina, Miff climbed into a tree and began to feed on the leaves of a vine, and Merlin swung to and fro above his mother and brother, hitting towards Pepe's head with one hand. Presently he dropped down, landing on his brother's shoulder, and with a wide play-face pulled at Pepe's hand. Idly Pepe tickled the infant whilst he continued to groom Marina with the other hand. Encouraged, Merlin became more vigorous, standing upright and pulling at the hair on Pepe's shoulders, crawling on to him, pushing between Pepe and his mother.

A casual observer passing by would, without doubt, have fancied that he saw a typical family group – a mother with her two young children and her mate. Only because we had watched these chimpanzees over a number of years did we know that, in fact, Pepe was Marina's son. In chimpanzee communities, of course, family groups comprise only a mother and some or all of her offspring; the father, apart from his necessary contribution to the conception of a child, plays no further part in its development. Indeed, neither we nor the chimpanzees normally have any idea as to which male was responsible for siring which child.

This exclusion of the male from familial responsibilities is, perhaps, one of the major differences between human and chimpanzee societies. For most human family groups look upon the father not only as the begetter of the children, but as the protector; and usually as the provider of food, or land, or money. Admittedly things are changing to-day in certain parts of the world where women are demanding equality and where free love results in many unmarried mothers – but these things are only happening in a very small minority of cultures if the peoples of the world are considered as a whole.

Human families, of course, vary enormously in structure. The smallest unit, the husband, wife and children, can be extended to include two or two hundred wives and any number of blood relations and relations by marriage. As yet we do not know whether the chimpanzee family group ever expands to include grandchildren as an integral part of the unit: certainly, though, it can never include the 'wife' or children of a male offspring or the 'husband' of a female offspring.

Despite this basic difference in the structure of the human and the chimpanzee family group, the behaviour of many human males is not so different from chimpanzee males as might be expected. In the Western world at any rate, many fathers, even though they may be materially responsible for their families' welfare, spend much time away from their wives and children – often in the company of other men. All-male groups are popular in many cultures: they range from clubs and stag-parties in the Western world, to initiation and warrior groups in more primitive societies.

In short, a vast number of human males, whilst they may be only too anxious at times for feminine company, are equally keen for much of the time to get away from women and relax in the ease of male companionship. Chimpanzee males seem to feel rather the same. Of course they cluster round pink females when these are available. But often they travel about and feed in all-male groups, and they are more likely to groom each other than they are to groom females or youngsters.

Never, however, have we seen anything which could be regarded as homosexuality in chimpanzees. Certainly a male may mount another in moments of stress or excitement, clasping the other round the waist, and he may even make thrusting movements of the pelvis – but there is no intromission. It is true, too, that a male may try to calm himself or another male by reaching out to touch or pat the other's genitals, but, whilst we still have much to learn about this type of behaviour, it certainly does not imply homosexuality. He only does this in moments of stress, and he will touch or pat a female on her genitals in exactly the same contexts. Figan, indeed, frequently touches his own scrotum in moments of sudden apprehension.

What about the normal heterosexual relationships which may develop between humans and those that may be observed between chimpanzees? The obvious difference between the two species lies in the fact that men and women are capable of establishing and maintaining monogamous relationships, both physical and spiritual, of long duration, and this sort of bond is unknown in chimpanzees. Monogamy, however, is far from being the only relationship found between men and women. In fact, human males, the world over, tend to be promiscuous. In some cultures this is accepted and men are permitted or even expected to have more than one wife. In other societies, where monogomy is the rule, it is nevertheless an accepted fact that unmarried – and even married – males will indulge in love affairs, or pick up women for a night, or visit brothels. Many young girls too will show promiscuous sexual behaviour if given the chance. It may, in fact, be that what we think of as true love – an emotion which embraces both the body and the mind of the beloved, which mellows with time and brings about harmony of living, which removes any need, in the man or the woman concerned, for another sexual partner – is, indeed, one of the rarest of human heterosexual relationships.

Sexual relationships between male and female chimpanzees are, in large part, similar to those which can be observed amongst many young people in England and America to-day. In other words, chimpanzees are very promiscuous. But this does not mean that every female will accept every male who courts her.

One young female, a little older than Fifi, showed a marked objection to the advances of the aggressive male Humphrey. Gigi invariably had a large male retinue when she went pink – indeed, from the first time she showed a true adult swelling it seemed that she had as much sex-appeal as Flo. But she simply could not bear Humphrey. When all the other males were satisfied, there would be Humphrey, his hair on end, glaring at Gigi, shaking branches, hunching his shoulders, stamping with his foot on the ground, moving cautiously towards her. All the while Gigi would be screaming and moving away from him. Sometimes Humphrey gave chase, but, though he once shook her out of the tree in which she had sought refuge, we never saw him actually 'rape' her. Quite often, though, he managed to get his way through dogged persistence. He went on and on courting her every time she went pink. His persistence was certainly rewarded eventually for, two years later, Gigi seemed almost to prefer Humphrey to any other male.

Fifi avoided Humphrey too, though she was less frightened of him than Gigi had been, and merely walked calmly away when he began to court her. Once, indeed, we watched Fifi walking and cantering around a tree no less than fifteen times with Humphrey, all his hair bristling, pursuing her. He could easily have caught her; instead he finally charged away in a wild display of frustration, hurling huge rocks, stamping his feet, and vanishing completely from the scene. In the same way, other females avoid other males.

Sometimes a male chimpanzee will actually insist on an unwilling female accompanying him on his travels until he is no longer interested in her, or she manages to escape. Indeed, this sort of masculine assertion of power led to a number of strange females being introduced to the feeding area in the old days. Once Hugo and I saw J.B. bustling down the slope whilst behind him, peering nervously in our direction, I recognised one of the old mothers with her three-year-old infant. She had never been to camp before and quickly we humans hid in one of the tents. As the pair got closer the female stopped, staring. Suddenly J.B. looked round and saw that she was not following. Standing upright he seized a sapling and swayed it from side to side until, with a whimper, the female hurried towards him and reached to touch his side, submissively. Her infant climbed a tree where he remained throughout most of the proceedings.

J.B. set off again but once more, after a few steps, the newcomer stopped, terrified by our tents. Just as J.B. looked back she turned to flee; quickly the big male pounded in pursuit, leapt on to her back and attacked her, stamping with his feet. The female, screaming loudly, escaped, but she only ran a short way and then turned, hurried back to J.B., and crouched in submission until he reached out and patted her again and again on her head. Once more he moved on and this time she followed; but only for a few yards. However, when J.B. again swayed a sapling, her fear of him overcame her fear of the strange place to which he was leading her, and she hurried to his side. Finally she followed him right into the feeding clearing. To our surprise (for J.B. was normally possessive of his food) she was permitted to take a large share of his bananas.

The strangest part of the story lies in the fact that this female showed no sign of a sexual swelling. For three days J.B. forced her to accompany him into camp, and then he appeared without her. When, some twelve days later, I saw the same female, fully pink, together with a whole retinue of male suitors, J.B. showed no more interest in her than any of the others. This sort of thing happens time and again, although some males only force sexually attractive females to follow them.

Leakey and Mr Worzle, more than any other males, have constantly forced different females to accompany them for days on end. For a long time both of them picked on nervous Olly – again often when she had no sign of a swelling. We never saw both of them with her at the same time – it was either one or the other.

When Fifi had been swelling regularly for about a year, and when most of the adult males were less interested in her than they had been when her pink state was a novelty, Leakey picked on her. Sometimes it seemed that Fifi was not at all loath to follow the old male: when she wanted to escape she usually edged away, cautiously, whilst Leakey's attention was elsewhere, but hurried back to him instantly if he spotted her manœuvre – before he had time, as it were, to get angry with her.

One day we shall never forget. Leakey, at that time, had become most peculiarly preoccupied with females. Constantly he forced first one and then another to accompany him. On this occasion he had just lost his current victim and was sitting eating bananas in camp when Fifi arrived on the scene. She was pink and, instantly, his bananas forgotten, he stood up, his hair bristling, and shook branches at her. She ran quickly up to him and presented. He mated her and then the two of them sat grooming each other. Suddenly Leakey looked up and saw Olly approaching camp. At once all his hair stood on end again and he began waving branches at *her*. Olly hurried up and Leakey began to groom her. Fifi, looking innocent, walked very slowly away. But Leakey noticed, his hair rose once more, and Fifi ran back pant-grunting in submission. Then Leakey began to try to get both his females to follow, but neither wanted to go with him. First he glared and shook branches at one until she ran up, and then at the other.

And so it went on until the strain and tension seemed suddenly to overwhelm Leakey – even as Fifi obediently approached him he ran at her and attacked her, rolling her over and over on the ground. At this Olly, of course, made a silent and rapid getaway and was soon lost to sight. The attack over, Leakey stood, his hair still on end, puffing from the exertion, whilst Fifi, crouching to the ground, screamed and screamed.

When Leakey noticed that Olly had gone he moved rapidly some way up the slope, peered round, ran to the other side of the clearing and looked round from there. And so, of course, Fifi escaped. Not for ten minutes or more did Leakey's hair slowly sink as he gave up the search and finally settled down to eat some bananas.

The relationship which the large Rodolf struck up with Flo, and which I described earlier, was, of course, rather different. Rodolf showed none of the bullying aggressive behaviour towards Flo which characterised the relationships of Leakey and the others to the females of their choice. Rodolf followed Flo wherever she went, and it was to him that she most often turned for comfort when she was hurt or upset. Also, Rodolf remained with Flo and her family for two weeks after her swelling had vanished.

It is, of course, fruitless to speculate as to the sort of heterosexual relationships which might develop if chimpanzee physiology were different: if, for example, Flo had been able to offer Rodolf continuing sexual satisfaction, if the female reproductive cycle of the chimp were the same as that of the human. The fact remains that female chimpanzees have evolved in such a way that they are only sexually receptive to males for a mere ten days per month; provided, that is, that they are neither pregnant nor lactating which, in older females, means that they may be denied sexual activity for up to five years. Along with this limited quota of sexual opportunities, evolution has landed them with ungainly sexual swellings.

How many times, while I have watched a hugely swollen female chimpanzee adjusting her position again and again, trying to get comfortable on some branch or hard rock, have I thanked evolution for sparing human females a similar periodic disfigurement – though the designers and manufacturers of bustles would, I suppose, have been in clover. Why, in heaven's name, has the female chimpanzee been so encumbered? At times the answer seems simple. One day I was sitting with Goliath and David Greybeard who were peacefully grooming each other. Suddenly Goliath stared intently across the valley and seconds later David followed suit. Even I, with my naked eye, soon spotted what looked like a large pink flower gleaming in a thickly foliaged tree. Already both males were off, moving fast through the undergrowth. I knew I could not keep up with them so I stayed where I was and presently watched both Goliath and David swing up into the tree, swagger around in the branches, and mate with the female.

In cases such as this the swelling, quite obviously, acts as a signal; and this, of course, can be important in a community where the

Olly's infant was the first victim of the polio epidemic (See p. 195)

Sniff adopted his sister after their mother's death (See p. 209)

Flo with Flint and Flame (*Photo P. McGinnis*)

With Grub in the forest

members split up so freely and where even females often wander about on their own. This is particularly so if one realises that all chimp females are not so forward as Fifi. Olly, indeed, sometimes seemed to try and hide away from males completely during her pink days. But there is a snag in this theory. If female chimpanzees developed conspicuous pink bottoms in order to signal to prospective suitors over a distance, then why did female baboons develop swellings? For they show enlargements quite as obvious as those of chimpanzees, and most baboons live in close groups where females are seldom out of sight of the males.

There are other arguments which could be put forward, but since none of them can be applied equally to chimps, baboons and the few other monkeys who show sexual swellings, there is little point in discussing them. To confuse the issue further, orang-utans, who live in even more widely scattered units and thicker forests than most chimps, do not develop swellings at all. The reason for swellings must, for the moment, perhaps for ever, remain something of a mystery.

Young female chimpanzees (even those with small infants) cycle much more regularly than old females. Flo and Olly, for instance, did not develop swellings for five years after the respective births of Flint and Gilka; nor did they show swellings during the early months of pregnancy. Melissa and other young females, on the other hand, not only cycled throughout four or more months of their eight-month pregnancies, but also started to show swellings again when their infants were only just over one year old. And it is primarily in young females that we see indications of quite stable preferences for individual males, and vice versa.

Figan was faithful to Pooch for over half a year – by which I mean not that he ignored other pink females but that, every time Pooch went pink, for six months running, the two went off together. So far as we know they were on their own at such times; certainly they did not visit camp where, without question, Pooch would have met other males.

Once I saw them going off together on the third day of Pooch's swelling. I followed. When they were about a quarter of a mile from camp they climbed into a tree and groomed each other for over an hour. Then they wandered on, fed for a while and, as dusk fell, climbed into another tree and made nests quite close to each other. I did not see them mating at all during that time. The next morning, when I arrived at their sleeping-tree at dawn, they mated once before wandering off peacefully together to feed.

Six days later Pooch, her swelling shrivelled, arrived at the feeding area on her own. About half an hour later Figan appeared, coming from the same direction. But they went off on their separate ways. We noticed that they never returned from such 'honeymoons' together – it was as though they didn't want to be found out. In reality, of course, the reason for the separate return probably lay in the fact that they only started to travel towards camp when Figan was no longer sexually interested in Pooch and so, on the way, they gradually drifted apart.

We shall never know how that relationship would have matured for, unfortunately, Pooch died when she was about ten years old. Since then Figan has often gone off in this way with Melissa. They wander the forests together, with Goblin accompanying them, and actual mating, as opposed to grooming and just being near each other, has only been observed comparatively infrequently. Interestingly enough, Faben also shows the same preference for Melissa and, recently, Melissa and Goblin spent most of their time first with one brother and then the other.

However, although such relationships may be shadowy forerunners of human love affairs, I cannot conceive chimpanzees developing emotions, one for the other, comparable in any way to the tenderness, the protectiveness, the tolerance and the spiritual exhilaration which are the hallmarks of human love in its truest and deepest sense. For chimpanzees usually show a lack of consideration for each other's feelings which, in some ways, may represent the deepest part of the gulf between them and us. For the male and female chimpanzee there can be no exquisite awareness of each other's bodies – let alone each other's minds. The most the female can expect of her suitor is a brief courtship display, a sexual contact lasting, at most, half a minute, and, sometimes, a session of social grooming afterwards. Not for them the romance, the mystery, the boundless joys of human love.

When I first began studying chimpanzees I often wondered whether a male and a female ever slept together at night. One evening I watched as a young female carefully tucked in the small leafy twigs that stuck out around her nest. Then she lay down. It was just light enough for me to see her lying there, about fifty yards away. A few moments later, however, she sat up and then, to my surprise, left her nest, climbed higher up the tree, and presently appeared at the edge of Mr McGregor's nest. The old male sat up and began to groom her. Five minutes later, though, he turned away and lay down and she left him to return to her own bed. When she finally settled down it was almost pitch dark – but soon the moon would be up.

As I hastened down the mountain to camp I determined to return, later on, and spend the night with the chimps. I had supper, wrote out my notes for the day and then, at about eleven o'clock, set off once more. The mountains and valleys were very quiet and still under the full moon and, in the coolness of night, I felt, as usual, that I could run up the steep slopes. When I reached the Peak I decided to have some coffee before starting my vigil, so I gathered a few twigs, lit a small fire, and hung my kettle on its waiting chain. Then I took my mug of coffee to the rocky perch overlooking the home valley. Below me the moonlight was reflected from the myriads of leaves that formed the upper canopy of the forest, glistening brightly on the smooth, shiny green of the palm fronds. Somewhere down there a baboon barked twice and then was silent. Behind me the darkness of the forest was closer; it was easy to imagine a leopard slinking through the trees, a herd of buffalo browsing the dank undergrowth.

I drank my coffee slowly, overawed by the beauty around me. The moonlight was so bright that only the most brilliant stars were shining, and the grey mist of the sky clung around the mountain peaks and spilled down into the valley below. It was a perfect night for love – for human love; but when I clambered down to the place from where I had watched the chimpanzees earlier, I could see, through binoculars, that the two chimps were still respectfully separate. Mr McGregor was flat on his back. The female was lying on her side, facing away from me, and I could just make out the pale gleam of her shell-pink swelling.

At about four o'clock the moon sunk behind the mountains of the Congo on the far side of the lake and, presently, even its lingering glow in the sky had faded. Then the night seemed very different – inky black and sinister with rustlings and crackling twigs all round. About an hour later the baboon troop in the valley below began to bark loudly and, after a few moments, the chimpanzees joined in with their fierce *waa* barks and loud pant-hoots. I guessed there was a leopard prowling along the valley and pulled the blanket closer round me. Now the night seemed less romantic, but I should have been just as glad to have Hugo's masculinity close beside me.

Presently the greyness of dawn filtered through the sky and the everyday world began to emerge. There was no movement from the chimpanzee's tree, but as it got lighter I saw that the female and Mr McGregor were still in their separate nests. At about six-fifteen the old male turned over, sat up, and then suddenly swung from his nest and, moving rapidly through the tree, leapt right into the female's bed. She, almost certainly woken from a deep sleep, uttered a piercing shriek and leapt out of her nest. Mr McGregor rushed after her, down through the wildly swaying branches, to the ground. Gradually her screams grew fainter as she fled through the forest, her impetuous suitor in hot pursuit. Not a very romantic ending to the night, perhaps, but it certainly showed that the female was very much in old McGregor's mind when he woke up that morning.

It was some years later, when Figan and Melissa were off on one of their trips, that, just before darkness fell, Figan moved from the branch where he had been sitting, chewing a wadge, and went towards Melissa's nest. The foliage was too thick to watch his progress, but there were no further sounds: none of the snappings of branches and twigs and the rustling of leaves that usually accompanies the making of a nest. In the morning, after the chimpanzees had left, we peered from as many vantage points as possible to try to determine how many nests there were. The tangle of branches and palm fronds was very thick, and it is quite possible that we missed something, but none of us saw more than one, and we all saw the same one – in the place where Melissa had made her bed the evening before. Had Figan and Melissa, perhaps, slept curled up together, with little Goblin to keep them company?

## Chapter 16 Baboons and Predation

Figan sat in the shade and groomed himself, and about ten other chimps groomed or lay resting in small groups that were scattered around the camp clearing. One adult male baboon was in camp also, cracking open palm nut kernels with his powerful teeth, and a juvenile baboon fed on the ripe fruit of one of the camp palm trees.

When Figan got up and began to walk towards this tree something in his gait, some tenseness in his posture, caused Mike to look towards him intently. Briefly, Figan glanced up at the young baboon and then, very slowly, began to climb the trunk. As he got higher the baboon peered down at him and started to make little shrill sounds, showing its teeth in a frightened grin. Then it jumped across to the crown of a neighbouring palm. Figan paused for a moment when he got to the top of his tree and then, very slowly, moved after the baboon. Screeching louder now, the youngster jumped back to its original tree. Figan followed. Twice more, still moving slowly, Figan went after his quarry from the original to the second tree and back again. And then, quite suddenly, he rushed towards the baboon which, terrified, took a huge leap into a tree some twenty feet below.

By this time the chimpanzees below were all intensely interested in the chase and stood gazing up, many of them with their hair on end. As the baboon leapt so Mike rushed up the tree towards it – at the same moment the adult male baboon came racing towards the scene uttering loud roaring sounds. The quarry, screaming loudly, took a second leap to the ground. Mike jumped down and chased after it, the male baboon chased Mike and, in the ensuing confusion, with the other chimpanzees milling around, the intended victim escaped.

Four years before that, when Figan had been a young adolescent,

Hugo and I had watched him creeping towards another juvenile baboon in a large fig tree. It had been Rodolf, so far as we could tell, who had actually initiated that hunt; he had walked towards the tree and stood, his hair very slightly on end and, if he had looked at the baboon at all, we had not noticed. Yet, as though at a signal, the chimpanzees who had been resting and grooming peacefully on the ground had got up and stationed themselves close to trees that would act as escape routes for the intended victim. And Figan, the youngest adolescent male of the group, had crept towards the baboon.

On that occasion too the prey had escaped; in response to its calls the whole baboon troop had rushed some hundred yards towards the scene and chimps and baboons had engaged in fierce conflict. Despite the fact that they had leapt at each other and called loudly, so far as we could see no physical damage had been inflicted on members of either side, but the young baboon had made its getaway in the confusion.

During the ten years that have passed since I began work at the Gombe Stream we have recorded chimpanzees feeding on the young of bushbucks, bushpigs and baboons, as well as both young and small adult red colobus monkeys, redtail monkeys and blue monkeys. And there are two authentic cases on record of chimpanzees in the area actually taking off African babies – presumably as prey, for one infant, when recovered from an adult male chimpanzee, had had its limbs partially eaten. Luckily these incidents occurred before I ever set foot at the Gombe Stream – in other words, before I had 'tamed' the chimpanzees.

Many people are horrified when they hear that a chimpanzee might eat a human baby, but after all, so far as the chimpanzee is concerned, men are only another kind of primate, not so very different from baboons in *their* eyes. Surely it should be equally horrifying to reflect on the fact that in a great many places throughout their range chimpanzees are considered a delicacy by humans?

Until recently, though we saw chimpanzees eating meat fairly frequently, those occasions when we could watch their hunting techniques were few and far between. During the past two years, however, with more people working at the Gombe, and with the chimpanzees so tolerant of their human observers, we have been able to learn a good deal about their way of hunting. Sometimes it appears that the capture of a prey is almost accidental; the chimpanzee, as he wanders along, stumbles across a baby bushpig, grabs it and the kill is made. On other occasions, however, the hunting seems to be a much more deliberate, purposeful activity, and often at such times the different individuals of a chimpanzee group show quite remarkable co-operation – as when different chimpanzees station themselves at the base of trees offering escape routes to a cornered victim.

I myself have only seen the actual killing of a prey animal on two occasions: once on that far-off day when Hugo and I watched a red colobus monkey seized and torn to pieces, and once, four years later, when a juvenile baboon was caught on the outskirts of camp. That was much more spectacular.

It happened one morning when Rodolf, Mr McGregor, Humphrey and an adolescent male were sitting, replete with bananas, and the baboon troop was passing through the camp area. All at once Rodolf got up and moved rapidly behind one of the buildings, followed by the other three. They all walked with the same silent, purposeful, almost stealthy pace that Figan had shown as he approached the palm tree that harboured his intended prey.

I followed – but even so I was too late to observe the actual capture. As I rounded the building I heard the sudden screaming of a baboon and then, a few seconds afterwards, the roaring of male baboons and the screaming and barking of chimpanzees. I ran the last few yards and, through some thick bushes, glimpsed Rodolf standing upright as he swung the body of a juvenile baboon above him by one of its legs and slammed its head down on to some rocks. Whether or not that was the actual death blow I could not tell: certainly the victim was dead as Rodolf, carrying it in one hand, set off rapidly up the slope.

The other chimps followed him closely, still screaming, and a number of adult male baboons continued to harass Rodolf, lunging towards him and roaring. This lasted only for a few minutes and then, to my surprise, they gave up. Presently the four chimpanzees appeared, climbing into the higher branches of a tall tree where Rodolf settled down and began to feed, tearing into the tender flesh of the belly and groin of his prey.

Other chimpanzees in the valley, attracted by the loud screaming and calling which typifies a hunt and kill, soon appeared in the tree, and a group of high-ranking males clustered around Rodolf, begging for a share of his kill. Often I have watched chimpanzees begging for meat and usually a male who has a reasonably large portion permits at least some of the group to share with him. Rodolf, however, protected his kill jealously that day. When Mike stretched out his hand, palm upwards in the begging gesture, Rodolf pushed it away. When Goliath reached a hand to his mouth, begging for the meat and leaf wadge which Rodolf was chewing, Rodolf turned his back. When J.B. gingerly took hold of part of the carcass, Rodolf gave soft threat barks, raised his arm, and jerked the meat away. And when old Mr McGregor reached up and tentatively took hold of the end of a dangling piece of gut, it was only by pure luck that yards and yards of intestine fell into his eager hands and festooned his bald crown and shoulders. Rodolf looked down and pulled at his lost food; the gut broke, and McGregor hurried to a distant part of the tree where a group of females and youngsters quickly gathered round to beg titbits from him.

Chimpanzees nearly always eat meat slowly, usually chewing leaves with each new mouthful as though to savour the taste for as long as possible. All in all, Rodolf kept almost the entire carcass to himself for nine hours that day, although, from time to time, he spat a wadge into a begging hand, or one of the other males managed to grab a piece from the carcass and make off with it. Sometimes, too, little scraps fell and then the youngsters were down the tree like a flash to search for the fragments in the undergrowth below. Often, too, I saw them actually licking the branches of the tree where the kill had touched them or below where, presumably, drops of blood had fallen.

At this time, the third year of Mike's supremacy, Rodolf was no longer the very high-ranking male he had been when first we knew him. How was it, then, that he dared push away Mike's hand, he who normally went into a frenzy of submission when Mike approached him? How dared he threaten Goliath and J.B. who were normally dominant to him? Even more puzzling – why did not these higher-ranking males snatch at least a part of the kill away from Rodolf? I had seen this sort of apparent inconsistency before during meat-eating episodes and I often wondered whether, perhaps, the chimps were showing the crude beginnings of a sense of moral values. Rodolf killed the baboon; the meat, therefore, is

Rodolf's. But more serious consideration of the behaviour has led me to suspect something rather different may be involved.

Mike would have attacked Rodolf without hesitation had the prize been a pile of bananas, yet, if Rodolf had gathered the fruits from a box for himself, they would have been his property quite as legitimately as was the meat. I wonder, then, if the principle involved may be similar to that which ensures that a territorial animal, within his own territory, is more aggressive, more likely to fight off an intruder, than if he met the same animal outside his territorial boundary. Meat is a much-liked, much-prized food item. An adult male in possession of such a prize may become more willing to fight for it and, therefore, less nervous of his superiors, than if he has a pile of everyday fruits like bananas. Indeed, in support of this theory, I should emphasise that in the early days of feeding, when bananas themselves were something of a novelty, the chimps very seldom *did* fight over the fruits.

And what of the lack of aggression in the more dominant males? Possibly, when they fail to detect those signals of apprehension which normally characterise their interactions with subordinates and, instead, meet very definite aggressive signals, they may feel hesitant in asserting their normal prerogatives.

This may be true of sexual behaviour too – it may be the reason why adult males do not normally fight for the privilege of mating with an attractive female but, instead, await their turn. Perhaps the male about to mount may, likewise, be prepared to defend his sexual right to the act of reproduction. And it is quite possible that the human sense of moral values may, in itself, have originated from basic behavioural patterns of this sort.

Be this as it may, we have, on many occasions, seen Rodolf and others guarding their meat from chimpanzees normally their social superiors. At such times the higher-ranking males, frustrated beyond endurance, frequently vent their aggression on lower-ranking individuals. And so, during the early part of a meat-eating session, before the big males have acquired portions for themselves, females and youngsters, as well as males of lower rank, are frequently chased quite violently through the branches, particularly if they venture too close to the kill.

It was in this context, in the old days, that the male chimpanzees were often extremely aggressive towards me. One day when

Rodolf was in possession of a kill and keeping everything to himself, J.B. hurried off to camp to snatch a few bananas. It was a journey of about two hundred yards and, on his return, he must have thought that, during his absence, I had managed to get a share of the carcass. He bustled through the thick undergrowth towards the meat-eaters, suddenly noticed me, stopped and stared. Slowly his hair began to bristle and all at once, with a loud bark, he charged straight towards me. I could not even try to get out of his way since the tangle of vines around was too thick and, as he stopped less than two feet from me and gave me a hard wham, I think I closed my eyes. But he merely picked up first my sweater and then my haversack, with quick frenzied movements, and sniffed them both intently. Then, setting them down, he turned and hurried off to try his luck with the real kill.

Another time it was worse. Hugo and I had been watching a large group of chimpanzees in a tree. Mike had a baboon kill and, whilst he was freely sharing with J.B. and a few others, five of the adult males had not succeeded in getting more than the odd scrap. These males kept displaying through the tree, chasing off any subordinates who happened to be in their way. The individual who appeared, to us, to be the most frustrated was David Greybeard.

Presently Mike swung from the tree and sat feeding in some thick vegetation on the ground. The other chimps, of course, followed him. Hugo and I, as we crept closer to watch what was going on, made the mistake of moving too quietly; as we suddenly came into their view, one of the young females, a relative newcomer to our group, took fright and rushed off. This startled all the others and there was a brief stampede. Then the chimps realised that it was a false alarm and, after glaring at us for a few moments, the five frustrated males charged towards us, running upright, waving their arms, and uttering loud and fierce-sounding threat barks. Just before they reached us they stopped – all except for David.

David Greybeard, as I have said earlier, could be a very aggressive chimpanzee indeed if sufficiently roused and, on that occasion, he most certainly was. Hugo and I, at the same moment, turned and fled. When we looked round, David was still coming. All at once Hugo, who was protecting me from the rear and struggling with his camera equipment, got caught in some vicious thorns. As he struggled to get free the enraged David still ran on, by then only a couple of yards from Hugo. Then, at the very last minute, he stopped and, with a final *waa* and upward jerk of his arm, turned and hurried back towards the other chimps.

Even to this day I feel sure that that episode was the closest shave we have had with any chimp. For David, despite his normal gentle disposition, was, in a way, the most dangerous so far as we were concerned because of his total lack of fear of humans. Many people have evinced surprise that Hugo and I ran, but this should not really be surprising. After all, the most submissive thing that a subordinate chimpanzee can do is to remove himself from the aggressor's presence – in other words, to run away. A social inferior who shows submissive behaviour such as crouching, presenting or holding out a hand towards an angry superior is quite likely to be attacked, whereas for the most part adult males do not persist in an aggressive incident if their intended victim runs off.

These days it seems that the chimpanzees not only have a far greater tolerance of the close proximity of humans, but they seem to realise that, when they have meat, their human watchers are unlikely to try to steal a share. Aggressive behaviour during meat-eating directed towards researchers at Gombe is becoming increasingly rare.

Just as hunting behaviour is interesting in that the chimpanzees show the beginning of co-operative endeavour so characteristic of human hunting societies, so is the consumption of meat fascinating because, normally, the possessor of the carcass is willing to share the meat with others – a characteristic not recorded for other non-human primate species in the wild. Often we have seen chimpanzees actually break off portions of their meat and hand them to begging individuals. One such incident was particularly remarkable.

It was a sudden outburst of chimpanzee and baboon calling that attracted us to the spot, and there we found Goliath with the freshly killed body of a baboon infant. Old Mr Worzle was up in the tree too, begging for a share and, as he did so, whimpering like an infant. Every few minutes Goliath got up and leapt away from him, but Worzle followed, crying, and when Goliath stopped, reached out to beg again. Suddenly, when Goliath pushed his hand away for at least the tenth time, Worzle threw a tantrum worthy of an infant, hurling himself backwards off the branch, screaming, and hitting the surrounding vegetation. Then, to our astonishment, Goliath, with a mighty tearing of the victim's skin and sinews, pulled the body of his prey in two and handed Mr Worzle the entire hind quarters. It was as though he could no longer endure the screaming and the commotion when he wanted to enjoy his prize. And Goliath shared with Flo too, when she presently arrived on the scene.

Mike, once he had become well established in his top-ranking position, became increasingly tolerant and benign during interactions with his subordinates. When he was in possession of a carcass he usually permitted the other males to share at least some parts of it with him. Once, when Mike had the remains of a baboon, I watched as he and Rodolf chewed together, one at each end of the carcass. Then the dominant male began to pull on his half. I had the impression not that he was trying to take away Rodolf's share, but simply that he wanted to tear the carcass in two. Never have I been more impressed by Rodolf's strength than on that occasion for, no matter how hard Mike pulled, standing up and heaving, Rodolf, as he held on to his end, simply remained almost motionless in his original position. He might have been solid rock for all the impression Mike's tugging and jerking made on his large frame. Presently Rodolf leant forward and, whilst Mike continued to pull, gnawed away at the skin until the carcass split in two, leaving Mike with the head.

The brain appears to be a special delicacy, and Mike very often ends up with the head of the prey. I have watched him gradually enlarging the *foramen magnum*, where the vertebral column joins the skull, biting away fragments of bone with his teeth. When the opening was large enough, he scooped out the brain with a crooked index finger. Sometimes the chimps open the brain by breaking away the frontal bones.

Once, when David was watching Mike eating brain and begging very persistently, Mike started to tickle him, as a mother will do with her child to distract it from some desired objective. After a few moments, the two adult males were both uttering loud chuckles of chimp laughter as they tickled each other. And it worked; when David could endure the tickling no more he moved away and Mike continued to eat brain undisturbed.

Only in one instance did we see Mike lose the prized head, with the brain intact, to another chimp. It was evening, and Mike had eaten well. As the last lurid rays of the sinking sun penetrated the forest canopy he held up the head of his victim, a juvenile baboon, and began to groom the fur. It was a gruesome sight, for one eye-socket was empty and from the other the eye dangled on its optic nerve. Just then there was a rustling in the undergrowth and J.B. emerged at a fast run, seized the head and vanished with it into the undergrowth. Mike, it seemed, was too full to care, for he did not pursue his friend. J.B. climbed into a nearby tree, made a nest, and sat there in his bed, eating mouthfuls of brain and leaf until darkness fell.

The picture which is emerging, from the years of information we have collected at Gombe, suggests that meat-eating behaviour in the chimpanzees occurs in cycles. It may be that a chance hunting success, when, for instance, a chimpanzee accidentally comes across a baby bushpig in the undergrowth, starts off a craze for meat in the group as a whole. While this craze lasts, for one or even two months, the adult and adolescent males may set out to hunt deliberately. Then, either because their craving is satisfied, or because a succession of hunts results in failure and they lose interest, the chimpanzees return to a diet of fruit, vegetables and insects. Until, a few months later, something triggers off another meat-eating craze.

I remember once when Humphrey came into camp with the skin of a red colobus monkey. When he left he was still clutching his trophy. About an hour later McGregor arrived, from the same direction as had Humphrey. Whether or not he had been able to get a share of that kill, he certainly had a craving for meat. About fifteen minutes after his arrival he suddenly stared, very intently, over the valley. Following his gaze we saw a small troop of red colobus monkeys. Almost at once McGregor set off with that fast, silent purposeful walk – and the other chimps in camp followed him. We got out the telescope and waited to see what would happen.

Presently there was a violent swaying of branches, monkeys began to chatter, and we glimpsed McGregor, as well as a few other chimps and a number of colobus, leaping through the trees. I saw a colobus chase a small chimp who leapt, screaming, towards the ground. Then, to our astonishment, we saw old McGregor pounding down an open grassy slope with a single male colobus bounding after him. It just shows that when an animal is really enraged it can appear far more terrifying than it really is – for that monkey would have come off very much the worse if Gregor had turned and attacked.

This observation makes it all the more surprising that adult male baboons, many times heavier and normally far more aggressive than male red colobus, are not more efficient as protectors when chimpanzees hunt their young. They do usually rush up and mob the chimpanzees, lunging at them, roaring, and generally making a commotion, when they notice that a hunt is going on, but so far, though the aggression looks quite fierce, we have never seen a chimpanzee hurt as a result of such an encounter. Once a male baboon actually leapt on to Mike's back and clung there for a few moments during a hunt, but Mike appeared none the worse afterwards. Even when the chimps have caught an infant baboon which is alive and screaming in their clutches, still the adult males do not attack. It is one more of those unsolved problems which makes the continuation of our research a challenge.

In fact, the whole question of the relationship between chimpanzees and baboons at the Gombe Stream is complex and very fascinating. In the early days of my study I found that, for the most part, individuals of the two species tended to ignore one another except for the youngsters who often played together. Sometimes adult chimps and adult baboons fed quite peacefully in the same tree although, if there were only a couple of chimps, they usually became restless and, finally, left when baboon after baboon climbed up into the branches of their tree. I also saw a number of aggressive incidents between groups of the two species; in the light of our knowledge to-day I suspect that these were attempts, on the part of the chimps, to catch young baboons.

Since 1963 when our feeding area came into being, interactions between chimpanzees and baboons have increased enormously in frequency: during the worst years, just before we built the feeding bunker, individuals of both species were hanging around camp for hours at a time in close proximity. At the beginning it seemed that, often, the baboons showed greater respect for the chimpanzees than they do to-day, but I think, in fact, this was simply because the baboons were more scared of humans then, so that only a few dared to move out into the camp clearing. Also we tried to discourage them by shining a mirror into their eyes, so that they were apprehensive anyway, and the sudden leaping, arm-waving threat of a chimpanzee was more likely to be effective. Even in those days, however, some male baboons were very aggressive indeed, and able to scare most of the chimps.

Subsequently, when students started following the baboons around, studying their behaviour, they lost most of their apprehension of humans and invaded our feeding area in ever larger numbers. Small wonder then that they seemed more aggressive, and that most of the chimps showed increasing respect for them. But whilst we watched many fights when baboon and chimp came into contact, we never saw serious wounds inflicted. For the most part, battles over bananas were based on threat and bluff, with the baboons roaring, lunging forward with open mouths or hitting towards their opponents, and the chimps leaping up, waving their arms, uttering loud *waa* barks or screaming.

Victory in such an aggressive conflict depends, to a large extent, on the individuals concerned. The chimps themselves quickly learned which of the male baboons could be chased off by bluff and which would stand their ground and should, therefore, be avoided. And the baboons likewise realised that female and young chimps were, for the most part, easy game; that some males such as David Greybeard could easily be intimidated into dropping a whole armful of bananas; that others, such as Goliath and Mike, were tougher customers. The baboons showed, too, a good deal of respect for old Mr Worzle – maybe they disliked his strange human-like eyes. Principally, though, it was because Worzle was usually fearless and simply stood his ground. Also he was the first chimpanzee we ever saw throwing stones and other objects at baboons who approached and threatened him. True, Worzle sometimes hurled leaves if no more suitable missile was close to hand, and once he threw a whole handful of bananas at an aggressive male baboon – to the obvious satisfaction of his opponent! As time went on, however, Worzle became more selective, usually picking up and throwing quite large rocks.

We were not surprised when first we saw Worzle throwing a stone at a baboon for we had seen chimps throwing things aggressively before. Rodolf, for instance, had hurled a large rock straight at Hugo and me the first time he ever visited camp. He had, it seemed, followed Goliath along without really looking where he was going. When he suddenly woke to the fact that there was a tent ahead, with us inside, he uttered a strangled call, stood upright, and threw his missile before bolting back to the bushes.

Soon after we first saw Mr Worzle throwing at baboons a number of the other males started doing likewise. These days nearly all the adult males use objects as weapons in this way but, though they usually choose quite large stones, they seldom hit the baboons unless they are at very close quarters.

At this point I must emphasise that, except during the actual distribution of bananas, relationships between chimpanzees and baboons, even at the feeding area, are normally peaceful, with individuals of both species showing a good deal of tolerance of each other. Indeed, it has often surprised us that, immediately after a frenzy of aggressive behaviour, chimpanzees and baboons, when all bananas are finished, will often sit close to each other as they relax and rest.

Since both chimpanzees and baboons are well known for their intelligence, it is not really surprising that, to some extent, individuals of the two species are able to communicate with each other. One day, for instance, a female baboon passed very close to Mr Worzle and seemed to startle him slightly. He raised his arm and gave a soft threat bark at which she instantly crouched and presented submissively. Mr Worzle then reached his hand towards her rump and almost certainly touched her in reassurance. At any rate, her posture became relaxed and she sat quite close to him. We have seen many other incidents of this sort.

When Flint was about eight months old he frequently tottered towards female baboons with sexual swellings and, to our initial surprise, these females often turned and presented their rumps to the small infant – just as a female chimpanzee would do. Some of them actually allowed him to touch their pink bottoms. Then we found that the same thing happened when Goblin, or any other chimpanzee infant, approached female baboons – though some of them were more tolerant than others in this context.

Particularly fascinating was the relationship between an old male baboon, known as Job, and many of our chimpanzees. Perhaps because of his age, perhaps for some other reason, Job took to

The orphan Merlin remained stunted and grew neurotic (see p. 205)

hanging about in camp, often leaning up against a tree-trunk as though life was altogether too much for him. One day we were amazed to see him approach Fifi and, turning sideways, signal his desire to be groomed. We were even more surprised when Fifi complied. After working on him for a few moments she, in turn, presented her side to him – but Job did not respond, and after a short while Fifi groomed him again for a few moments before walking away. After this it became quite common to see Job going up to different young chimps, begging them, by his posture, to groom him and, fairly often, being briefly groomed in response.

On another occasion Figan walked up to the old male baboon with that jaunty, slightly swaggering walk which signified his intentions were playful. He reached out and gave Job a few prods and then repeatedly chucked him under the chin. Finally, when none of these activities called forth any sort of response from the baboon who simply sat looking, if anything, somewhat bewildered, Figan tried a new tactic. He pressed his forehead against Job's and, time and time again, butted him so that the baboon's whole body was shaken. After a moment or two of this treatment Job, it seemed, could tolerate no more, and he made a threatening movement, lunging slightly towards Figan with yellowed teeth showing. Figan did not appear to be unduly perturbed by this but, nevertheless, after a second or two he walked away. We saw two other adolescent male chimps trying to play with Job, thumping him and seeming to tickle him with their fingers, but they never succeeded in sparking off any sign of playful behaviour in the old male.

Despite all these friendly interactions, and the many times we have watched young chimps and baboons playing together, we have not observed another friendship like that described earlier between Gilka and the young female baboon Goblina. True, there was one very playful juvenile male baboon who quite often indulged in wild games with Flint, and even a sub-adult male who, for a while, played frequently with Fifi when she was about six. But these relationships did not persist more than a few weeks, and the individuals concerned played only when chance threw them together. The friendship between Gilka and Goblina, was unique in that it lasted for nearly a year and, quite definitely, the two sought each other's company.

Ultimately they got used to my presence

One day, when chimpanzee screams and baboon roars took Hugo and me running towards a tall tree, we saw two adult male chimps with the newly killed body of an infant baboon. We were so busy watching the behaviour of the chimps that for a while we paid no attention to the mother baboon who was repeatedly running at the chimps, calling loudly. But when we did see who it was we could hardly bear to watch any longer, for it was Goblina – the chimps had killed her first infant.

After about thirty minutes Goblina moved away, with the one male baboon who had remained with her at the scene of the kill. But later she returned, still with the same young male, and sat farther up the slope, staring towards the chimpanzees. Every so often she uttered a low-pitched grunt that, to Hugo and me, sounded like a call of distress. She stayed there for ten minutes or so and then once more went away. However, during the four hours that the chimps remained in the area, she returned three times more, by herself. Three hours later she went back yet again to the deserted scene of the crime, still alone. And, every few minutes, she uttered that desolate-sounding call.

A year later Gilka's old playmate had a second infant which, to-day, is a juvenile and, with luck, will not fall victim to the predatory attack of a chimpanzee. But the baby that we hoped would become a proper playmate for Gilka, her own sibling, had a fate possibly worse than that of Goblina's first infant.

## Chapter 17  Death

Olly's new baby was four weeks old when he suddenly became ill. I had been excited when I heard of his birth: would his elder sister Gilka show the same fascination for him as Fifi had for Flint? And how would Olly react if she did? Though I was not at the Gombe when the baby was born, I was there a month later when, one evening, Olly walked slowly into camp supporting him with one hand. Each time she made a sudden movement he uttered a loud squawk, as though in pain, and he was gripping badly. First one hand or foot and then another slipped from Olly's hair and dangled down.

Whilst Olly sat eating her bananas, Gilka groomed her mother. Often I had watched Gilka working her way closer and closer to her small sibling's hands just as Fifi had done, two years earlier, when Flint was tiny. Previously Olly had always pushed Gilka's hands away but this time she permitted her daughter actually to groom the baby's head and back.

Next morning it was obvious that the baby was very ill indeed. All his four limbs hung limply down and he screamed almost every time his mother took a step. When Olly sat down, very carefully arranging his legs so as not to crush them, Gilka went and sat close to her mother and stared at the infant. But she did not even attempt to touch him.

Olly ate a couple of bananas and then set off along the valley, with Gilka and me following. Olly only moved for a few yards at a time and then, as though worried by the screams of her infant, sat down to cradle him close. When he quietened she moved on again, but, of course, he instantly began to call out so that, once more, she sat to comfort him. After travelling about a hundred yards, which took her just over half an hour, Olly climbed into a tree. Again she carefully arranged her baby's limp arms and legs on her lap as she sat down. Gilka, who had followed her mother, stared again at her small sibling, and then mother and daughter began to groom each other. The baby stopped screaming and, apart from occasionally grooming his head briefly, Olly paid him no further attention.

When we had been there some fifteen minutes it began to pour, a blinding deluge which almost obscured the chimps from my sight. During that storm, which went on for thirty minutes, the baby must either have died or lost consciousness; when Olly left the tree afterwards he made no sound and his head lolled back as limply as his arms and legs.

I was amazed at the sudden and complete change in Olly's handling of her baby. I had watched a young and inexperienced mother carrying her dead baby and, even the day after its death, she had held the body as though it were still alive, cradling it against her breast. But Olly climbed down the tree with her infant carelessly in one hand and, when she reached the ground, she flung the limp body over her shoulder. It was as though she knew he was dead. Perhaps it was because he no longer moved or cried that her maternal instincts were no longer roused.

The following day Olly arrived in camp, followed by Gilka, with the corpse of her infant slung over her shoulder. When she sat down the body sometimes dropped heavily to the ground. Occasionally Olly pushed it into her groin as she sat; when she stood she held it by an arm or even a leg. It was gruesome to watch, and several of the young female chimpanzees went over and stared. So also did a number of baboons. Olly ignored them all.

Presently Olly wandered away from camp and she and Gilka, with me following, went some way up the opposite mountain slope. Olly seemed dazed; she looked neither to right nor to left, but plodded up the narrow trail through the forest, the body slung over her neck, until she reached a place half-way up the mountainside. Then she sat down. The dead infant slumped to the ground beside her and, other than to glance down briefly, Olly ignored it. She just sat, staring into space, hardly moving for the next half-hour save to hit away the fast-gathering swarm of flies.

Now, at last, came Gilka's opportunity to play with her sibling. It was not easy to watch. Already the corpse had begun to smell; the face and belly showed a definite greenish tinge, and the eyes, which were wide open, stared glassily ahead. Inch by inch, glancing repeatedly up at her mother's face, Gilka pulled the body towards her. Carefully she groomed it, and then she even tried to play, pulling one dead hand into the ticklish spot between her collar bone and neck, and actually showing the vestige of a play-face. We had been so glad for Gilka's sake when old Olly had given birth again – but it seemed that Gilka was always to be ill-fated. Presently, with another quick glance towards her mother, Gilka carefully lifted the dead body of her sibling and pressed it to her breast. Only then did Olly's lethargy leave her for a moment. She snatched the body away but then, once more, let it fall to the ground.

When the old female moved on, it was merely to plod down the same track and so back to camp. She ate two bananas, sat staring into space, and then wandered off up the valley.

For another three hours I managed to keep up with the little family. Every ten minutes or so Olly just sat or lay, and Gilka, as before, groomed or tried to play with the dead body of her sibling. Finally Olly became worried by my presence and started to walk fast, glancing at me over her shoulder. She headed into a dense thicket, and though I continued after her a short way, I presently decided I should desist. In fact, I was glad to get out of the tangled vines for in the hot and humid air the smell of death was heavy where Olly had passed, and each twig had its own quota of bloated flies.

The following afternoon Olly and Gilka arrived in camp without the body. Somewhere in the valley Olly must finally have abandoned it.

Had we known, at the time, that Olly's infant was, without doubt, the first victim of the terrible paralytic disease that struck our chimpanzee community, I should never have followed the family – for, at that time, my own baby was on the way. But we had no suspicion, and the next victims did not appear for another two weeks. Later we discovered that there had been a bad outbreak of poliomyelitis amongst the African population in the Kigoma district; since chimpanzees are susceptible to almost every human infectious disease, and are known to get polio, it seems almost certain that the epidemic which afflicted our chimpanzees was, in fact, polio. We traced two of the human victims to a village ten miles to the south of our feeding area – just outside the Park, but in a valley where chimpanzees had been seen on many occasions. Possibly it was there that the first chimpanzee had contracted polio and, from that place, the disease had spread northward to affect those in our group.

When we realised that the disease was probably polio, we panicked, for neither Hugo nor I nor Alice, our student assistant, had received a full course of polio vaccine. We got through to Nairobi on the radio telephone and spoke to Louis. He arranged for a plane to fly down to Kigoma bringing sufficient vaccine for ourselves, our African staff – and the chimpanzees. We did not know to what lengths the epidemic might ravage the chimp community and we felt it was worth at least trying to stop it by treating those that were healthy.

The Pfizer Laboratories in Nairobi generously supplied us with the oral vaccine, and we offered it to the chimps in bananas. Each individual was supposed to have three drops once a month for three months. Most of them took the doctored fruit unhesitatingly, but a few of them spat the banana out after one mouthful – despite the fact that, to us, the drops were quite tasteless. For these fussy individuals we had to prepare three bananas, each with one drop of vaccine, instead of putting three drops in one fruit. We had to be careful, too, that a high-ranking individual, who had had his monthly dose, was not around when we dosed a subordinate from whom he might have snatched the banana – and thus got a double dose.

I think those few months were the worst I have ever lived through for, every time a chimp stopped visiting the feeding area for a while, we started to wonder whether we would ever see him again. Or, worse, if he would reappear hideously crippled. Fifteen chimpanzees, in our group, were afflicted, of whom six lost their lives. Some of the victims were lucky and survived with only minor disabilities; Gilka lost partial use of one hand, and Melissa was affected in her neck and shoulders. The magnificent young males, Pepe and Faben, both appeared after short absences trailing one useless arm. One adolescent male returned, after a long absence, shuffling along in a squatting position and with both arms paralysed. He could only eat the bits and pieces he was able to reach with his lips and he was nothing but a skeleton covered with dull staring hair. We had to shoot him. And there were other victims, like fat, bustling J.B., of whom we had all become so fond, who just disappeared, and we could only conjecture about their lonely deaths. But it is the nightmare of Mr McGregor's illness that still haunts us in the wakeful hours of the night.

It was quite late in the evening when Hugo noticed Flo, Fifi and Flint moving cautiously towards a low bush, just below camp, staring intently and, every so often, uttering soft worried calls as they stood upright to peer over the long grass. We hurried down to see what was happening. We saw the flies first. Every leaf and twig near the bush bore its burden of metallic blue and green flies, buzzing angrily as our approach disturbed them. As we cautiously moved closer we expected to see some dead creature – but it was Mr McGregor, and he was still alive. He was sitting on the ground reaching for the tiny purple berries that grew on the bush above his head, stuffing them into his mouth. It was not until he wanted to reach another cluster of the fruit that we realised the horror of what had happened. Looking towards the berries, the old male seized hold of a low branch and pulled himself along the ground – both his legs trailed uselessly after him. When next he wanted to shift his position he put both hands behind him on the ground and inched his body backwards in a sitting position.

Flo and her family soon moved away, but Hugo and I stayed there until darkness fell. To our amazement Mr McGregor was able to pull himself up into a low-branched tree, using only his powerful arms. He hauled himself quite high and then managed to build a small nest. As he climbed we saw the reason for the hoard of flies, for he had lost the use of the sphincter muscle of his bladder and, every time he strained to reach a higher branch, a spurt of urine trickled down his paralysed thighs. We saw, too, that his bottom was raw and bleeding, and realised that he must have dragged himself a long way in his efforts to return to camp. Indeed, the following morning we followed his trail, and the narrow lane of flattened vegetation led us some fifty yards downhill to the stream and continued twice as far again on the opposite side before we lost it in a steep gulley on the mountain-side.

The next ten days – and they seemed more like ten years – had a nightmare quality. We kept hoping to notice some flicker of life return to his paralysed legs, but he never twitched as much as a toe. During this time, he did not move from the vicinity of our feeding area. In the mornings he usually remained in his nest until eleven o'clock or even later. Then, slowly, he lowered himself to the ground, where he sat for half an hour or so, just looking round and occasionally grooming himself. After this he usually dragged himself for a few yards to some low-growing food and ate for a while.

We found that he had another way of moving from place to place – a laborious kind of somersaulting, head over heels. When we first saw this we were overjoyed for we felt sure that there must be some muscles still active in his thighs – but soon we realised that it was merely the incredible strength in his arms that enabled him to raise his body from the ground together with the dead weight of his paralysed legs. He could only move in this way when there were tough grass tufts or tree roots that he could grip with his hands for leverage.

Usually he was back in bed by half-past four or so and, during his whole illness, he only made use of three different nests, two of which were in one tree. Once, near the beginning of it all, we watched as he started to climb three different trees – each one proving impossible for him in his handicapped state. After the tremendous effort of lifting himself to the lower branches he had to lower himself laboriously to the ground again.

Of course we nurtured him. At first he was apprehensive if we approached too closely and threatened us with a quick raising of one arm and a soft bark. But after two days he seemed to sense that we were trying to help – and after this he even lay back and allowed me to pour water from a sponge into his open mouth. We made a little basket of leaves which we filled with food – bananas, palm nuts, any wild foods we could collect – and pushed it up to him in his nest on the end of a long stick. When he had vacated his nest in the mornings, we climbed up and cleaned it for him – because, of course, he could no longer move to defecate over the side.

We soon realised that Mr McGregor was being driven to distraction by the huge swarm of flies, so we took an aerosol spray whenever we visited him and squirted all round – killing, each time, well over a thousand of the revolting bloated things. At first, this too scared the old chimp, but he very soon seemed to understand what it was all about and then he welcomed the operation.

One of the most tragic things about the whole tragic affair was the reaction of the other chimps to the stricken old male. Initially, almost certainly, they were frightened by the strangeness of his condition. We had noticed the same thing when some of the other polio victims appeared in camp for the first time. When Pepe, for instance, shuffled up the slope to the feeding area, squatting on his haunches with his useless arm trailing behind him, the group of chimps already in camp stared for a moment and then, with wide grins of fear, rushed for reassurance to embrace and pat each other, still staring at the unfortunate cripple. Pepe, who obviously had no idea that he himself was the object of their fear, showed an even wider grin of fright as he repeatedly turned to look over his shoulder along the path behind him – trying to find out, presumably, what it was that was making his companions so frightened. Eventually the others calmed down, but, though they continued to stare at him from time to time, none of them went near him – and presently he shuffled off, once more on his own. Gradually the other chimps got used to Pepe, and soon the muscles in his legs became strong enough to enable him to walk about upright – as had Faben from the start.

But McGregor's condition was, of course, far worse. Not only was he forced to move about in an abnormal manner, but there was the smell of urine and the bleeding rump and the swarm of flies buzzing around him. The first morning of his return to camp, as he sat in the long grass below the feeding area, the adult males, one after the other, approached with their hair on end and, after staring, began to display around him. Goliath actually attacked the stricken old male who, powerless to flee or defend himself in any way, could only cower down, his face split by a hideous grin of terror, whilst Goliath pounded on his back. When another adult male bore down on McGregor, hair bristling, huge branch flailing the ground, Hugo and I went to stand in front of the cripple and, to our relief, the displaying male turned aside.

After two or three days the others got used to McGregor's strange appearance and grotesque movements, but they kept well away from him. There was one afternoon that, without doubt, was from my point of view the most painful of the whole ten days. A group of eight chimps had gathered and were grooming each other in a tree about sixty yards from where McGregor lay in his nest. The sick male stared towards them, occasionally giving little grunts.

Mutual grooming normally takes up a good deal of a chimpanzee's time, and Mr McGregor had been drastically starved of this important social contact since his illness.

Finally he dragged himself from his nest, lowered himself to the ground and, in short stages, began the long journey to join the others. When at last he reached the tree he rested for a while in the shade and then made the final effort and pulled himself up until he was close to two of the grooming males. With a loud grunt of pleasure he reached a hand towards them in greeting – but even before he made contact they both swung quickly away and, without a backward glance, started grooming on the far side of the tree. For a full two minutes old Gregor sat motionless, staring after them. And then he laboriously lowered himself to the ground. As I watched him sitting there alone, my vision blurred, and when I looked up at the groomers in the tree I came nearer to hating a chimpanzee than I have ever done before or since.

For several years Hugo and I had suspected that the aggressive adult male Humphrey was, in fact, Mr McGregor's younger brother. The two travelled about together frequently and often the older male had hurried to Humphrey's assistance when he was being threatened or attacked by other chimps. But it was during the last days of Mr McGregor's life that we became convinced that these two males were siblings: no bond, other than that of a family, could have accounted for Humphrey's behaviour both then and afterwards.

During Gregor's illness Humphrey seldom moved farther than a few hundred yards away from the old male – although even he never actually groomed him. Sometimes Humphrey went away across the valley to feed, but within an hour or so he was back, resting or grooming himself near his paralysed friend. On the first day of his return to camp Gregor climbed quite high in a tree and made a nest. Suddenly Goliath began to display around him, swaying the branches more and more vigorously, slashing the old male on the head and back. Gregor's screams grew louder, and he clung to the rocking branches tightly. Finally, as if in desperation, he let himself drop down through the tree, from branch to branch, until he landed on the ground. Then he started to drag himself slowly away. And Humphrey, who for as long as I had known him had been extremely nervous of Goliath, actually leapt up into the tree, displaying wildly at the much higher-ranking male and, for a brief moment, attacking him. I could hardly believe it.

One day Mr McGregor managed to pull himself right to the feeding area, up thirty yards of very steep slope, to join a large number of chimpanzees who were eating there. We were able to give him a whole box to himself so that, for a while at least, he was part of the group again. When the others moved away up the valley, Gregor tried to follow. But whether he dragged himself on his belly, or hitched himself backwards, or laboriously somersaulted, he could move only very slowly, and the rest of the group was soon out of sight.

Then, after about five minutes, we saw Humphrey coming back. For a few moments he stood and watched Gregor's progress, then he turned and moved after the other chimps again. But once more he came back and stood waiting for the old male. This time he actually shook grasses at Mr McGregor, as though he were trying to force a reluctant female to follow him. Eventually Humphrey abandoned his attempts to follow the big group and built his nest close to Gregor's just below the observation area.

On the tenth evening, when we went down with his supper, Mr McGregor was not in his nest, nor could we see him sitting in the grass. When we found him, after a short search, we soon realised that, somehow, he had dislocated one arm. And then we knew that, in the morning, we should have to shoot our old friend. We had known it, secretly, all along – yet we had waited, hoping for a miracle. I stayed with him for a while and, as dusk fell, he looked up more and more often into the tree above him. I realised that he must want to make a nest, so I cut and took to him a large pile of green vegetation. At once he manœuvred himself on to it, lay down and, with one hand and his chin, tucked the twigs over to make a comfortable pillow.

I went down to see him later that night, and it says much for the extent to which we had won his trust and confidence that, having heard my voice, he closed his eyes and went back to sleep, three feet away and with his back to me and my bright pressure lamp. Next morning, whilst he was grunting in delight over his favourite food – two eggs which we had given him – we sent him, unsuspecting, to happier hunting grounds.

We did not allow any of the chimps to see his dead body, and it seemed that, for a long time, Humphrey did not realise that he would never meet his old friend again. For nearly six months he returned again and again to the place where Gregor had spent the last days of his life, and sat up some tree or other, staring around, waiting, listening. During this time he but seldom joined the other chimps when they left together for a distant valley; he sometimes went a short way with such a group but, within a few hours, he usually came back and sat, staring over the valley, waiting, surely, to see old Gregor again, listening for the deep almost braying voice, so similar to his own, that was silenced for ever.

## Chapter 18  Mother and Child

Five-year-old Merlin had been among the first victims of the polio epidemic. Though he had been one of our favourite youngsters, playful and impish, we were almost glad when he died for, by that time, he had become a pathetic wreck of a chimpanzee, emaciated, lethargic and morose. But I must go back to the beginning and relate the whole story.

When Merlin was about three years old, still suckling, still riding about on his old mother Marina, and sleeping with her at night, the two of them stopped visiting the feeding area. Merlin's six-year-old sister, Miff, continued to appear regularly – and since, before, she had always travelled about with her mother and small brother, we presumed that Marina and Merlin had died. Then, just over three months later, Merlin reappeared, following his eldest brother, thirteen-year-old Pepe, into camp. He looked thin, with a tight hard belly, and his eyes seemed enormous as though he had not slept for a long time. Goodness knows what had happened to his mother, or how long she had been dead – she must have died for we never saw her again.

It appeared that the chimpanzees who were at the feeding area at the time of Merlin's return had not seen him for a long time, for they hurried to greet him, embracing and kissing and patting the infant. He ate some bananas and then sat, huddled up, close to his big brother. Later on that morning Miff arrived – at once she hurried over to Merlin and brother and sister began to groom each other. Merlin only groomed Miff briefly, but she groomed him rapidly and intently for more than fifteen minutes. When Miff was ready to go she looked back over her shoulder towards Merlin and waited until he followed before going on: just like a mother waiting for her child.

From that moment Miff, to all intents and purposes, adopted her little brother. She waited for him when she went from place to place, she allowed him to share her nest at night, she groomed him as frequently as his mother would have done. For the first few days after his return she even permitted him to ride occasionally on her back, but, after that, she pushed him off if he tried to jump on – she was a thin, long-legged youngster and Merlin was probably too heavy for her. It seemed that Pepe was with his young siblings more than he had been when Marina had been alive – possibly Miff, deprived of a mother, followed him instead. And when he was there Pepe normally went to Merlin's assistance during moments of social excitement.

Gradually, as the weeks passed, Merlin became more emaciated, his eyes sank deeper into their sockets, and his hair grew dull and staring. He became increasingly lethargic and played less and less frequently with the other youngsters. In other ways, too, his behaviour began to change.

One day Merlin was sitting grooming with Miff when a group of chimpanzees approached along the forest path. Miff instantly got up and hurried some way up a tree as Humphrey, who was in the forefront of the group, started the pant-hoots which indicated the imminence of an arrival display. Two other females nearby also hurried out of the way, but Merlin began moving fast towards Humphrey, pant-grunting in submission. Humphrey, who had already started to display, ran straight at Merlin, seized him by one arm and dragged him for several yards along the ground. As the big male charged away Merlin, screaming, rushed to embrace Miff. He had behaved like a small infant who does not yet appreciate the signals of impending aggression in his elders. Yet before this, Merlin, like all normal three-year-olds, had always responded instantly and appropriately to signals of this sort.

This episode, in fact, was the start of a marked deterioration of Merlin's social responses. Time and again he was dragged or buffeted by displaying males because he ran towards them instead of away. When he was four years old Merlin was far more submissive than other youngsters of that age: constantly he approached adults to ingratiate himself, turning again and again to present his rump, or crouching, pant-grunting, before them. At the other end of the scale, Merlin was extra aggressive to other infants of his own age group: when Flint approached to try to play, Merlin, whilst he sometimes merely crouched or turned his back, was equally likely to hit out aggressively so that Flint, a year younger than he, ran off with squeaks of fear. And, matching this decline in playful behaviour, Merlin began to groom older chimps, particularly his sister, more frequently and for longer at a time than other youngsters of his age.

As Merlin entered his sixth year his behaviour was becoming rapidly more abnormal. Sometimes he hung upside down, like a bat, holding on to a branch by his feet and remaining suspended, almost motionless, for several minutes at a time. Often he sat, hunched up with his arms around his knees, rocking from side to side with wide open eyes that seemed to stare into the far distance. And he spent much time grooming himself during which he pulled out hair after hair, chewed at their roots, and dropped them.

One of the strangest observations concerned Merlin's tool-using behaviour during the termite season. I had often watched him, as a two-year-old, playing about near Marina whilst she fished for hours at a time after the manner of old females. Occasionally Merlin went to watch his mother intently, and once I saw him take up a thin twig and, holding it rather as a human infant first holds a spoon, prod with it at the surface of the termite heap.

The following year, just before his mother died, his tool-using ability was improved, though still far from efficient, as was only to be expected in a three-year-old. Nearly always he chose grasses or twigs that were too short – about two inches long as compared with the eight- to twelve-inch lengths favoured by adults – and he manipulated his tiny tools clumsily and incompetently, pushing them fairly carefully into termite passages but then almost instantly yanking them out – so that even if a termite had managed to cling on it would, undoubtedly, have been jerked off. Once he pushed a thick piece of straw firmly into a hole from which he was unable to extract it.

Unfortunately I was not at the Gombe Stream for the following termite season, and Merlin's tool-using behaviour was not recorded. But the next year, when he was five years old, I watched him on several occasions. Normally a youngster is efficient at this age, both in his choice and his manipulation of grass tools, so that I was amazed to find that Merlin's technique had hardly improved since I had watched him two years earlier. He still chose, for the most part, minute tools, and if he selected one that was of reasonable length, it was limp or bent. He still jerked his tools from the holes rather than withdrawing them with the care of an adult. It was indeed strange – particularly as Miff was a keen termite-fisher and Merlin must have spent many hours with her whilst she worked heap after heap.

Only in one respect had his behaviour matured. Two- and three-year-old chimp youngsters seldom spend more than a couple of minutes at a time working at the job. After this they move off and play around for a while before making another brief attempt. But Merlin worked away with the concentration of a chimpanzee older than his five years; on one occasion he persisted, without interruption, for forty-five minutes. But during that time he caught only one termite, and that was on the end of his finger when he was trying to enlarge a hole!

By this time Merlin was so thin that every bone showed. His hair was not only dull, but there were great patches of it missing on his legs and arms where he had gradually pulled it out during self-grooming. Often he lay stretched flat on the ground whilst the other youngsters played, as though he was constantly exhausted. He was definitely smaller than Flint who was more than a year his junior.

Just before the start of the short rains we wondered, for a while, whether he was beginning to improve, for he began to play again. Some of his games with Flint, Goblin and Pom were quite vigorous – but as soon as one of his playmates got rough, Merlin still either crouched, squealing in submission, or else turned and hit out aggressively. Despite this improvement, however, when the rains began we all felt convinced that he could not survive six months of cold and wet. At the slightest shower he started to shiver, and often his face and hands and feet actually went blue with cold. That is why we were, in many ways, relieved when polio put an end to his sufferings.

Merlin is not the only orphan we have known; three other infants lost their mothers and two of them, like Merlin, were adopted by their elder siblings. Beattle lost her mother when she was about the same age as Merlin had been when Marina died; but Beattle's sister was two or three years older than Miff and a much

208

In Hugo I found a kindred spirit
*Copyright National Geographic Society*

stronger female, bigger in build. Beattle not only travelled around with her big sister and slept with her at night, but also was allowed to ride about on her broad back.

Beattle showed similar signs of depression to those shown by Merlin: she too became rather emaciated, she too played less and less frequently. But, at about the time when Merlin's behaviour had begun to deteriorate even more, Beattle's began to improve. And during *her* sixth year Beattle behaved just like a normal child of that age – except that, as Merlin had been, she was still very dependent on her older sister. Unfortunately these two sisters did not visit camp frequently, and just after this they stopped coming altogether for months on end. Then the elder of the two appeared several times on her own. We still do not know whether Beattle managed to survive.

In some ways Sorema was the most tragic orphan for she was only just over a year old when her mother died and still almost completely dependent on her for transport, protection and, most important, food. For solid foods do not start to play an important role in the diet of an infant chimpanzee until it is over two years old.

During the two weeks that Sorema survived her mother she was carried everywhere by her six-year-old brother, Sniff. It was a touching sight to see the juvenile male moving about with his tiny sister, pressing her against his breast with one hand, cuddling and grooming her. When they were in camp Sorema ate a few bananas, but it was milk she needed, and each day she seemed weaker and her eyes looked bigger. Then, one morning, Sniff came into the feeding area cradling her dead body.

It seems strange that an orphaned infant should be adopted by an elder sibling rather than by an experienced female with a child of her own who could, perhaps, provide the motherless youngster with milk as well as with adequate social protection and security. Three-year-old Cindy was an only child when her mother died – and she was not adopted at all. Previously she and her mother had spent days at a stretch in the company of another mother and we fully expected that Cindy would continue to move about with this adult female, especially as we had always thought the two females were sisters. However, Cindy was never seen travelling with her mother's friend but moved about either alone or trailed along with

David Greybeard

any group she happened to encounter. She showed signs of depression quicker than the other three-year-old orphans, and less than two months after her mother's death she stopped visiting the feeding area and was never seen again.

Why does a three-year-old chimpanzee become so depressed when he loses his mother? True, he is still dependent to some extent on her milk – but he only suckles for a couple of minutes every two hours, and he is able to eat the same solid foods as an adult. We do not yet know the answer to this question, but we have a clue if we look at the differences in behaviour shown by Merlin and Beattle after their mothers had died.

Both of these infants had been deprived of their mothers at a similar age, and deprived of the reassurance of the breast. Both, initially, showed gradually increasing depression. But then Merlin's condition declined whereas Beattle's improved. Beattle was able to continue riding about on another chimpanzee, just as she had ridden her mother before her death. Her world, indeed, had been shattered by the loss of her mother; if her sister moved away only a few yards without her she whimpered, even screamed, and rushed after her. But once she had scrambled aboard she was, once more, in close physical contact with a large chimpanzee – an individual who knew what to do in times of trouble, who would rush her to safety up a tree at the right time, who could run fast and swiftly and carry them both to safety.

Merlin, in contrast, no longer had a haven of refuge after Marina's death. Miff was no more than a constant companion and was of little use to her brother in times of social excitement in the group. And so it seems possible that Merlin's troubles were principally psychological; that his terrible physical condition resulted more from a sense of social insecurity than from any nutritional deficiency caused by the absence of his mother's milk. This theory is to some extent supported by the fact that, when his physical condition was at its worst, just before he died, he did seem to cheer up a little mentally, as though, very slowly, his mind might have been recovering. But, by that time, it was too late.

If, one day, we are able to study the development of a chimpanzee orphan to adulthood, we may learn much. Will time heal the wound caused by the death of the mother? What abnormalities will persist as a result of his early traumatic experience? The answers may be beneficial to those studying orphaned or socially deprived human children. For, whilst chimpanzee society does impose certain rules of conduct, it imposes far fewer than even the most primitive human society. A human has amazing powers of self-control and he learns, early in life, what are the accepted norms of behaviour. This means that unless he is mentally unbalanced he is usually able to control, at least in public, any inclination he may have to behave in an unacceptable way. A chimpanzee, however, is not inhibited by any fear of 'making a fool of himself'.

We can, of course, learn much about the effect of an early traumatic experience in the subsequent adult life of a human, but not only is a man's behaviour much more complex, it is also more difficult to make consistent, regular observations on an adult human. So a real understanding of the less complicated behaviour of a chimpanzee orphan during successive years may prove invaluable to our better understanding of some of the problems faced by human orphans.

We hope that, at least, we will be able to make a thorough study of one chimpanzee whose behaviour has been quite severely disturbed – not by the loss of his mother, but by a change in her attitude towards him. This, together with some other circumstances, produced symptoms in the child not unlike those which characterised the behaviour of our orphans. I am referring to Flint.

Flint was just under five years old when Flo became pink for the first time since his birth. This time, unlike that marathon five-week swelling which heralded Flint's arrival, Flo was fully pink for only four or five days; during which Flint was not permitted to suckle. Flo fended him off and often played with him vigorously every time he tried. When her swelling shrivelled, though, he resumed suckling again – despite the fact that she was, as we subsequently discovered, pregnant. In addition, Flint continued to sleep with Flo at night and often to ride about on her back.

Flint's prolonged infancy was possibly due to Flo's extreme age; to the fact that she no longer had the strength to battle with her somewhat obstreperous child. For, in point of fact, she had started to try to wean him when he was three years old. But Flint, from the start, had shown remarkable persistence in getting access to the breast; on those occasions when Flo had not given in after a few moments to his gentle pushing and whimpering, Flint had flown into terrible tantrums, hurling himself about on the ground, flailing his arms, rushing down the slope screaming until Flo, after staring for a moment in the direction of his receding screams, had plodded in pursuit to reassure her son and suckle him. As Flint had grown older he had taken to hitting and biting his mother when she refused him the breast – and, whilst Flo had sometimes retaliated, she had, at the same time, held the child very close as though trying to reassure him even while she bit or cuffed him. After these violent episodes she had always given in and allowed him to suckle in the end.

When Flo had been pregnant for six months her milk seemed suddenly to dry up. Deprived of this comfort at last, Flint had gone through the same stage of babyish behaviour that had marked Fifi's weaning five years earlier: constantly he clung to his mother, whimpering if she set off only a few yards ahead of him, frequently pushing in when Flo was grooming one of her other offspring and crying if she did not instantly turn her attention to him.

We felt really sorry for old Flo during the final months of her pregnancy. It was the height of the dry season and swelteringly hot. Yet, as she moved slowly along the mountain tracks, she had to support not only the new baby within her hugely swollen belly, but the half-grown male body of Flint, perched ridiculously on her frail old body. Sometimes, when I followed them, I feared that Flo would be unable to survive the birth of yet another infant. Every few yards she stopped to rest, and her eyes often looked vacant and far away. Also Flint was a bully to his old mother. When she had sat for a few minutes he was anxious to go on, keen to reach the fruit trees for which they were heading. He would push and push at his mother's back, whimpering louder and louder, until she got up and once more plodded on, Flint instantly climbing on to her. Sometimes he actually kicked her if she did not respond to his pushing by hand. And, when he was not pestering her to move on, he was bothering her to groom him, whimpering and pulling at her hands if she ignored him.

We saw the new baby, Flame, when she was only just dry, early in the morning. It is possible that Flo moved to a new nest during the moonlit night, or even climbed down to give birth on the ground, for at dawn, she had not been in the nest which she had made the evening before.

Flame, even as a newborn baby, was pretty – unlike the ugly little Goblin. She had a pale face in which her eyes, misty with extreme youth, looked almost blue. We were amazed, though, at how little hair she had – her chest and belly, and the insides of her arms and legs, were completely pink and naked.

We followed Flo that morning as she wandered along, accompanied by Flint, supporting, with one hand, Flame and the placenta to which the baby was still attached. Somewhat to our surprise, Flint's behaviour was exemplary – as it was, indeed, for the first few weeks after Flame's birth. He stopped pestering Flo for grooming, and he no longer tried to ride on her back. He often went up to touch Flame, but when Flo gently pushed his hand away he did not persist.

I shall never forget when, after his attempts to make contact with his little sibling had been repeatedly prevented, Flint deliberately picked a small twig and reached out to touch Flame with *that*. Then he withdrew it and intently sniffed the end. This happened for the first time when Flame was less than a day old – afterwards we saw him do the same thing a number of times. He was, in fact, using the twig as a tool to aid him in his attempt to find out about the new baby.

Flame rapidly developed into a healthy, alert and somewhat precocious infant. Flint, however, after his initial good behaviour, began to show signs of reverting to the dependent, clinging and irritating child he had been before his sister's birth. Once more he demanded to climb on to Flo's back when the family was on the move: indeed, there were a few occasions, when Flint was suddenly scared, when he clung underneath, embracing Flame's tiny body as well as his mother's with his arms and legs. Also he threw the most terrible tantrums if Flo did not immediately permit him to climb with her and Flame into their communal bed.

As the weeks went by Flint's behaviour began, increasingly, to resemble that of an orphaned youngster: he declined more and more invitations to play with other youngsters; he spent longer and longer periods grooming; he became noticeably listless and lethargic. He did not, however, show any inclination to harm Flame, unlike some human children who are jealous of a new sister or brother. Indeed, when Flame was three months old and Flo less possessive of her, both Flint and Fifi spent much of their time playing with Flame, grooming or cuddling her. Sometimes these two took it in turns to carry Flame through the forest whilst Flo plodded along behind them, seemingly quite unconcerned.

When Flame was six months old Flo became very ill with the same flu-like disease which had claimed old William, and which has stricken the chimpanzee nearly every rainy season. For six days Flo, Flint and Flame were missing, and all the students at the Research Centre went out on search-parties, hunting for them. When she was finally found, Flo was so sick that she could hardly move – and Flame had gone. We expected that Flo would die for she could neither eat nor climb off the wet ground. But, amazingly, the tough old female regained her health and strength.

From this time onward Flint, who had not been ill at all, began to show a change in his behaviour. As the days went by he lost his lethargy and again began to play frequently with his old playmates, to whirl and pirouette and somersault as in the days before Flame's birth. But he did not relinquish his babyish behaviour – in fact, so far as dependency on Flo was concerned, Flint became more clinging than ever. For a while we thought that he might actually start to suckle again for he was constantly putting his lips to Flo's nipples: but it seemed that, during her illness, his mother's milk had dried up. In all other respects, however, Flint again became Flo's baby. She shared her food with him, she permitted him to climb on to her back, or even, on occasions, cling to her belly. She groomed him constantly and, as of old, she welcomed him into her bed at night. And this state of affairs persisted until Flint was over six years old.

What went wrong with Flint's upbringing? Had he, as a small infant, been 'spoilt' by too much attention from his mother, sister and two big brothers? With the backing of that close-knit family he certainly got his own way nearly all the time in his dealings with individuals outside the family group. Did Flo fail to wean her son at the proper time because she was so old that she could no longer cope with Flint's tantrums when she denied him access to the breast?

Whatever the reason for Flo's failure there can be no doubt but that Flint, to-day, is a very abnormal juvenile. Will he gradually lose his peculiarities as he grows older, or will some traces of infantile behaviour characterise him when he is mature? This question, like so many others, can only be answered by the continuation of our research at the Gombe Stream. Already, however, we have been repeatedly impressed by the extent to which the growing child depends on his mother. Who would have thought that a three-year-old chimpanzee might die if he lost his mother? Who would have guessed that at five years of age a child might still be suckling and sleeping with his mother at night? Who would have dreamed that a socially mature male of about eighteen years of age would still spend much time in the company of his old mother?

It seems, so far, that most wild chimpanzee mothers are quite efficient at rearing youngsters. Nevertheless, we have learned too that occasional maternal inadequacies – such as Flo's inability to wean Flint at the proper time or the somewhat callous attitude of Passion to her infant Pom – may have marked consequences for the youngsters concerned.

I was at the Gombe Stream for several months during 1966 when my own child was on the way and also during the following year when he was with me as a tiny baby. I watched the chimpanzee mothers coping with their infants with a new perspective. From the start Hugo and I had been impressed with many of their techniques and we made a deliberate resolve to apply some of these to the raising of our own child. Firstly we determined to give our baby a great deal of physical contact, affection and play. He was breast-fed, more or less on demand, for a year. He was not left to scream in his crib. Wherever we went we took him with us so that though his environment was often changing, his relationship with his parents remained stable. When we punished him we quickly gave him reassurance through physical contact and, when he was small, we tried to distract him rather than simply prevent him from doing something naughty.

As he grew older it became increasingly necessary, of course, to temper chimpanzee techniques with our own common sense – after all, we were dealing with a human, not a chimpanzee, infant. Nevertheless, we tried not to punish him for errors until he reached an age when he could understand the reason behind the reprimand, and we continued to keep him with us and to give him frequent physical and mental reassurance.

Has our method of bringing him up been successful? We cannot say as yet. We can only point out that to-day, at four years of age, he is obedient, extremely alert and lively, mixes well with other children and adults alike, is relatively fearless and thoughtful of others. In addition, and quite contrary to the predictions of many of our friends, he is very independent. But then, of course, he might have been like this anyway, even if we had brought him up in a quite different way.

*Chapter 19* In the Shadow of Man

The amazing success of man as a species (if success can indeed be said to be the proper word), is the result of the evolutionary development of his brain which has led, among other things, to tool-using and tool-making, the ability to solve problems by logical reasoning, thoughtful co-operation, and language. One of the most striking ways in which the chimpanzee, biologically, resembles man lies in the structure of his brain. The chimpanzee, with his marked capacity for primitive reasoning, exhibits a type of intelligence which is closer to that of man than is the case with any other mammal living today. The brain of the modern chimpanzee, in fact, is probably not too dissimilar to the brain that so many millions of years ago, directed the behaviour of the first ape-men.

Prior to that far-off day when first I watched David Greybeard and Goliath modifying grass stems so that they could use them to fish for termites, the fact that prehistoric man made tools was considered to be one of the major criteria which distinguished him from other creatures. As I pointed out earlier, the chimpanzee does not fashion his probes to 'a regular and set pattern' – but then prehistoric man, before his development of stone tools, undoubtedly poked around with sticks and straws. At that stage it seems unlikely that he made tools to a set pattern either.

It is because of the close association, in most people's minds, of tools with man, that special attention has always been focused upon any animal able to use an object as a tool; but it is important to realise that this ability, on its own, does not necessarily indicate any special intelligence in the creature concerned. The fact that the Galapagos woodpecker finch uses a cactus spine or twig to probe insects from crevices in the bark is, indeed, a fascinating phenomenon, but it does not make the bird more intelligent than a genuine woodpecker which uses its long beak and tongue for the same purpose.

The point at which tool-using and tool-making, as such, acquire evolutionary significance is, surely, when an animal can adapt its ability to manipulate objects to a wide variety of purposes, and when it can use an object spontaneously to solve a brand new problem which, without the use of a tool, would prove insoluble.

At the Gombe Stream alone we have seen chimpanzees use objects for many different purposes. They use stems and sticks to capture and eat insects and, if the material picked is not suitable, then it is modified. They use leaves to sop up water which they cannot reach with their lips – and first they chew on the leaves and thus increase their absorbency. One individual used a similar sponge to clean out the last smears of brain from the inside of a baboon skull. We have seen them use handfuls of leaves to wipe dirt from their bodies or to dab at wounds. They sometimes use sticks as levers to enlarge underground bees' nests.

So far no chimpanzee has succeeded in using one tool to make another. Even with tuition, one chimpanzee, the subject of exhaustive tests, was not able to use a stone hand-axe to break a piece of wood into splinters suitable for obtaining food from a narrow pipe. She could do this when the material was suitable for her to break off splinters with her teeth but, although she was shown how to use the hand-axe on tougher wood many times, she never even attempted to make use of it when trying to solve the problem.[1] However, many other chimpanzees must be tested before we say that the chimpanzee, as a species, is unable to perform this act. Some humans are mathematicians – others are not.

And so, when the performance of the chimpanzee in the field is compared with his actual abilities in test situations, it would seem that, in time, he might develop a more sophisticated tool culture. After all, primitive man continued to use his early stone tools for thousands of years, virtually without change. Then, suddenly, we find a more refined type of stone tool culture appearing, widespread across the continents. Possibly a stone-age genius invented the new culture and his fellows, who undoubtedly learned from and imitated each other, copied the new technique.

1 Khroustov, H. F. (1964) Formation and highest frontier of the implemental activity of anthropoids. XIIth Internat. Congress of Anthropological and Ethnological Sciences. Moscow.

If the chimpanzee is allowed to continue living he too might suddenly produce a race of chimp super-brains and evolve a brand new tool-culture. For it seems almost certain that, whilst the ability to manipulate objects is innate in a chimpanzee, the actual tool-using patterns practised by the Gombe Stream chimpanzees are learned by the infants from their elders. We saw one very good example of this. It happened when a female had diarrhoea: she picked a large handful of leaves and wiped her messy bottom. Her two-year-old infant watched her intently and then twice picked leaves and wiped his own clean bottom.

\*　　\*　　\*

To Hugo and me, and indeed to many scientists interested in human behaviour and evolution, one very significant aspect of chimpanzee behaviour lies in the close similarity of many of their communicatory gestures and postures to those of man himself. Not only are the actual positions and movements similar to our own, but also the contexts in which they often occur.

When a chimpanzee is suddenly frightened he frequently reaches to touch or embrace a chimpanzee close by, rather as a girl, watching a horror film, may seize her companion's hand. Both chimpanzees and humans seem reassured, in stressful situations, by physical contact with another individual. Once David Greybeard caught sight of his reflection in a mirror; terrified, he seized Fifi, then only three years old. Even such contact with a very small chimp seemed to reassure him; gradually he relaxed and the grin of fear left his face. Humans, indeed, may sometimes feel reassured by holding or stroking a dog, or some other pet, in moments of emotional crisis.

This comfort which the chimpanzee and human alike appear to derive from physical contact with another, probably originates during the years of infancy when, for so long, the touch of the mother, or the contact with her body, serves to calm the frights and soothe the anxieties of both ape and human infants. So, when the child grows older and his mother is not always close by, he seeks the next best thing – close physical contact with another individual. If his mother is around, however, he may deliberately pick her out as his comforter. Once when Figan was about eight years old, he was threatened by Mike. He screamed loudly and hurried past six or seven other chimps nearby until he reached Flo;

then he held his hand towards her and she held it with hers. Calmed, Figan stopped screaming almost at once. Young humans too continue to unburden their hearts to their mothers long after the days of childhood have passed; provided, of course, that an affectionate relationship exists between them.

There are some chimps who, far more than others, constantly seem to try to ingratiate themselves with their superiors. Melissa, for one, particularly when she was young, used to hurry up and lay her hand on the back or head of an adult male almost every time one passed anywhere near her. If he turned towards her, she often drew her lips back into a submissive grin as well. Presumably Melissa, like the other chimps who constantly seem to try to ingratiate themselves in this way, is simply ill at ease in the presence of a social superior so that she constantly seeks reassurance through physical contact. If the dominant individual touches her in return, so much the better.

There are, of course, many human Melissas: the sort of people who, when trying to be extra friendly, reach out to touch the person concerned and smile very frequently and attentively. Usually they are, for some reason or other, people who are unsure of themselves and slightly ill at ease in social contexts. And what about the smiling? There is much controversy as to how the human smile has evolved. But it seems fairly certain that we have two rather different kinds of smile, even if, a long time ago, they derived from the same facial gesture. We smile when we are amused and we smile when we are slightly nervous, on edge, apprehensive. Some people, when they are nervous at an interview, smile in this way at almost everything that is said to them. And this is the sort of smile that can probably be closely correlated with the grin of the submissive or frightened chimpanzee.

When chimpanzees are overjoyed by the sight of a large pile of bananas they pat and kiss and embrace one another rather as two Frenchmen may embrace when they hear good news, or as a child may leap to hug his mother when told of a special treat. We all know those feelings of intense excitement or happiness which cause people to shout and leap around, or to burst into tears. It is not surprising that chimpanzees, if they feel anything akin to this, should seek to calm themselves by embracing their companions.

I have already described how a chimpanzee, after being threatened or attacked by a superior, may follow the aggressor, screaming and crouching to the ground or holding out his hand. He is, in fact, begging a reassuring touch from the other. Sometimes he will not relax until he has been touched or patted, kissed or embraced; Figan several times flew into a tantrum when such contact was withheld, hurling himself about on the ground, his screams cramping in his throat, until the aggressor finally calmed him with a touch. I have seen a human child behaving in the same sort of way, following his mother around the house after she has told him off, crying, holding her skirts, until finally she picked him up and kissed and cuddled him in forgiveness. A kiss or embrace or some other gesture of endearment is an almost inevitable outcome once a matrimonial dispute has been resolved, and the clasping of hands to denote renewal of friendship and mutual forgiveness after a quarrel occurs in many cultures.

It is if we begin to consider the moral issues at stake when one human begs forgiveness from another, or himself forgives, that we get into difficulties when trying to draw parallels between human and chimpanzee behaviour. In chimpanzee society the principle involved when a subordinate seeks reassurance from a superior, or when a high-ranking individual calms another, is in no way concerned with the right or wrong of the aggressive act. A female who is attacked for no reason other than that she happens to be standing too close to a charging male is quite as likely to approach the male and beg a reassuring touch as is the female who is bowled over by a male as she attempts to take a fruit from his pile of bananas.

Again, whilst we may make a direct comparison between the effect, on anxious chimpanzee or human, of a touch or embrace of reassurance, the issue becomes complicated if we probe into the motivation which directs the gesture of the ape or the human who is doing the reassuring. For humans are capable of acting from purely unselfish motives; we can be genuinely sorry for someone and try to share in his troubles in an endeavour to offer comfort and solace. It is unlikely that a chimpanzee acts from feelings quite like these; I doubt whether even members of one family, united as they are by strong mutual affections, are ever motivated by pure altruism in their dealings one with another.

On the other hand, there may be parallels in some instances.

Most of us have experienced sensations of extreme discomfort and unease in the presence of an abject, weeping person. We may feel compelled to try to calm him, not because we are sorry for him, in the altruistic sense, but because his behaviour disturbs our own feeling of well-being. Perhaps the sight – and especially the sound – of a crouching, screaming subordinate similarly makes a chimpanzee uneasy; the most efficient way of changing the situation is for him to calm the other with a touch.

There is one more aspect to consider in relation to the whole concept of reassurance behaviour in chimpanzees, and that is the possible role played by social grooming in the evolution of the behaviour. For the chimpanzee – and indeed for many other animals too – social grooming is the most peaceful, most relaxing, most friendly form of physical contact. Infant chimpanzees are never starved for physical contact for they spend much time close to their mothers. Then, as they get older, they spend more time away from their mothers and also more time playing with other youngsters; and play, typically, involves a good deal of physical contact. As the youngster matures he gradually plays less frequently; at the same time he spends longer and longer socially grooming, either with his mother and siblings or, as he gets older, with other adults. Sometimes a grooming session between mature individuals may last for two hours. The obvious need for social grooming was well demonstrated when old Mr McGregor, with his paralysed legs, dragged himself those sixty long yards to try and join a group of grooming males.

When a chimpanzee solicits grooming he usually approaches the selected partner and stands squarely in front of him, either facing him with slightly bowed head or facing away and thus presenting his rump. Is it possible, then, that submissive presenting of the rump, and submissive bowing and crouching, may have derived from the postures used to solicit grooming? That in the mists of the past the subordinate approached his superior, after he had been threatened, to beg for the reassuring, calming touch of grooming fingers? If so, then the response of the chimpanzee thus approached, the touch or the pat, may equally have been derived from the grooming pattern. Indeed, on some occasions a few brief grooming movements do occur when a dominant individual reaches out in response to the submissive posture of a subordinate.

It is quite reasonable to suppose that such a response may have become ritualised over the centuries so that to-day the chimpanzee usually gives a mere token touch or pat in place of grooming his submissive companion.

\*     \*     \*

When two chimpanzees greet each other after a separation their behaviour often looks amazingly like that shown by two humans in the same context. Chimpanzees may bow or crouch to the ground, hold hands, kiss, embrace, touch or pat each other on almost any part of the body, particularly the head and face and genitals. A male may chuck a female or an infant under the chin. Humans, in many cultures, show one or more of these gestures. Even the touching or holding of another's genitals is a greeting in some societies; indeed, it is described in the Bible, only it has been translated as the placing of the hand under the companion's thigh.

In human societies much greeting behaviour has become ritualised. A man passing an acquaintance in the street does not necessarily incline his head to show that he acknowledges the superior social status of the other, yet undoubtedly the gesture derives from submissive bowing or prostration. We do not only smile when we are nervous and ill at ease during a greeting; nevertheless, our own greetings often serve to acknowledge the relative social status of the individuals concerned, particularly on formal occasions.

A greeting between two chimpanzees nearly always serves such a purpose – it re-establishes the dominance status of the one relative to the other. When nervous Olly greets Mike she may hold out her hand towards him, or bow to the ground, crouching submissively with down-bent head. She is, in effect, acknowledging Mike's superior rank. Mike may touch or pat or hold her hand, or touch her head, in response to her submission. A greeting between two chimps is usually more demonstrative when the individuals concerned are close friends, particularly when they have been separated for days rather than hours. Goliath often used to fling his arms around David, and the two would press their lips to each other's faces or necks when they met; whereas a greeting between Goliath and Mr Worzle seldom involved more than a casual touch even when the two had not seen each other for a while.

It is not only the submissive and reassuring gestures of the chimpanzee that so closely resemble our own. Many of his games are like those played by human children. The tickling movements of chimpanzee fingers during play are almost identical to our own. The chimpanzee's aggressive displays are not unlike some of ours. Like a man an angry chimpanzee may fixedly stare at his opponent. He may raise his forearm rapidly, jerk back his head a little, run towards his adversary upright and waving his arms, throw stones, wield sticks, hit, kick, bite, scratch and pull the hair of a victim.

In fact, if we survey the whole range of the postural and gestural communication signals of chimpanzees on the one hand and humans on the other, we find striking similarities in many instances. It would appear, then, that either man and chimp have evolved gestures and postures along a most remarkable parallel, or that we share, with the chimpanzees, an ancestor in the dim and very distant past; an ancestor, moreover, who communicated with his kind by means of kissing and embracing, touching and patting and holding hands.

One of the major differences between man and his closest living relative is, of course, that the chimpanzee has not developed the power of speech. Even the most intensive efforts to teach young chimps to talk have met with virtually no success. Verbal language does indeed represent a truly gigantic stride forward in man's evolution.

Chimpanzees do have a wide range of calls, and these certainly serve to convey some types of information. When a chimp finds good food he utters loud barks: other chimps in the vicinity instantly become aware of the food source and hurry to join in. An attacked chimpanzee screams, and this may alert his mother, or a friend, who may hurry to his aid. A chimpanzee confronted with an alarming and potentially dangerous situation utters his spine-chilling *wraaaa* – again, other chimps may hurry to the spot to see what is happening. A male chimpanzee, about to enter a valley or charge towards a food-source, utters loud pant-hoots – and other individuals realise not only that another member of the group is arriving but also which one. To our human ears each chimpanzee is characterised more by his pant-hoots than by any other type of call. This is significant since the pant-hoot, in particular, is the call which serves to maintain contact between the scattered groups of the community. Yet the chimps themselves can certainly recognise individuals by other calls – for instance, a mother knows the scream of her offspring. Probably a chimpanzee can recognise the calls of most of his acquaintances.

However, whilst chimpanzee calls do serve to convey basic information about some situations and individuals, they cannot, for the most part, be compared to a spoken language. Man, by means of words, can communicate abstract ideas; he can benefit from the experiences of others without having to be present at the time; he can make intelligent co-operative plans. All the same, when humans come to an exchange of emotional feelings, most people fall back on the old chimpanzee-type of gestural communication – the cheering pat, the embrace of exuberance, the clasp of hands. And when, on these occasions, we use words too, we often use them in rather the same way as a chimpanzee utters his calls – simply to convey the emotion we feel at that moment. 'I love you. I love you,' repeats the lover, again and again, as he strives to convey his overwhelming passion to his beloved – not by words but by his embraces and caresses. When we are surprised we utter inanities such as 'Golly!' or 'Gosh!' or 'Gee whizz!' When we are angry we may express ourselves with swear words and other more or less meaningless phrases. This usage of words, on the emotional level, is as different from oratory, from literature, from intelligent conversation, as are the grunts and hoots of chimpanzees.

Recently it has been proved that the chimpanzee is capable of communicating with people in quite a sophisticated manner. There are two scientists in America, Allen and Beatrice Gardner, who have trained a young chimpanzee in the use of the approved sign language of the deaf and dumb. The Gardners felt that, since gesture and posture formed such a significant aspect of *chimpanzee* communication patterns, such a sign language might be more appropriate than trying to teach vocal words.[1]

Washoe was brought up, from a tiny infant, constantly surrounded by human companions. These people, from the start, communicated not only with Washoe, but also with each other when in the chimp's presence, in sign language. The only sounds they made were those approximating chimpanzee calls such as laughter, exclamations and imitations of Washoe's own sounds.

1 Gardner, R. A. and B. T. *Science* Vol. 165, pp. 664-72.

Their experiment has been amazingly successful. At five years of age Washoe can understand some three hundred and fifty different symbols, many of which signify clusters of words rather than just a single word; and she can also use about one hundred and fifty of them correctly herself. The Gardners have been criticised for allowing Washoe to use 'sloppy' signing. In fact, some signs were taught to Washoe in a slightly different form when this corresponded quite closely with one of her own gestures. Other signs Washoe changed of her own accord, when she was small – and interestingly it transpired that many of these adaptations were exactly the same as those used by small human children. In other words, they represent the 'baby talk' of the deaf and dumb child. Many of these Washoe corrected as she grew older.

I have not seen Washoe, but I have seen some film demonstrating her level of performance, and, strangely enough, I was most impressed by an error she made. She was required to name, one after the other, a series of objects as they were withdrawn from a sack. She signed off the correct names very fast – but even so, it could be argued that an intelligent dog would ultimately learn to associate the sight of a bowl with a correct response of one scratch on the floor, a shoe with two scratches, and so on. And then a brush was shown to Washoe – and she made the sign for a comb. That to me was very significant – it is the sort of mistake a small child might make, calling a shoe a slipper, or a plate a saucer – but never calling a shoe a plate.

Perhaps one of the Gardners' most fascinating observations concerns the occasion when, for the first time, Washoe was asked (in sign language) 'Who is that?' as she was looking into a mirror. Washoe, who was very familiar with mirrors by that time, signalled back, 'Me, Washoe'.

That is, in a way, a scientific proof of a fact we have long known – that, in a somewhat hazy way perhaps, the chimpanzee has a primitive awareness of Self. Undoubtedly, there are people who would prefer not to believe this since, even more firmly rooted than the old idea that man is the only tool-making being, is the concept that man alone in the animal kingdom is Self conscious. Yet this should not be disturbing. It has come to me, quite recently, that it is only through a real understanding of the ways in which chimpanzees and men show similarities in behaviour that we can reflect, with meaning, on the ways in which men and chimpanzees *differ*. And only then can we really begin to appreciate, in a biological and spiritual manner, the full extent of man's uniqueness.

Man is aware of himself in a very different way from the dawning awareness of the chimpanzee. He is not just conscious that the body he sees in a mirror is 'I', that his hair and his toes belong to *him*, that if a certain event occurs *he* will be afraid, or pleased, or sad. Man's awareness of Self supersedes primitive awareness of a fleshly body. Man demands an explanation of the mystery of his being and the wonder of the world around him and the cosmos above him. So Man, for centuries, has worshipped a God; has dedicated himself to science; has tried to penetrate the mystery in the guise of the mystic. Man has an almost infinite capacity for preoccupation with things other than Self; he can sacrifice himself to an ideal, immerse himself in the joys or sorrows of another; love, deeply and unselfishly; create and appreciate beauty in many forms. It should not be surprising that a chimpanzee can recognise himself in a mirror. But what if a chimpanzee wept tears when he heard Bach thundering from a cathedral organ?

In his long quest for truth the scientist has never been able to provide a platform for Man's ancient belief in God and the spirit. Yet who, in the silence of the night or alone in the sunrise, has not experienced – just once perhaps – a flash of knowledge 'that passeth all understanding'? And for those of us who believe in the immortality of the spirit, how much richer life must be.

Yes, man indeed overshadows the chimpanzee. Yet the chimpanzee is, nevertheless, a creature of immense significance to the understanding of man. Just as he is overshadowed by us, so the chimpanzee overshadows all other animals. He has the ability to solve quite complex problems, he can use and make tools for a variety of purposes, his social structure and methods of communication with his fellows are elaborate, and he shows the beginnings of self-awareness. Who knows what the chimpanzee will be like forty million years hence? It should be of concern to us all that we permit him to live, that we at least give him the chance to evolve.

# Chapter 20  Man's Inhumanity

As the barbed arrow sank into her flesh, Flo lurched and staggered, clinging to the branch. Flint clung to her, screaming in fright, and the blood from his mother's wound slowly dropped down on to his face. As I watched, unable to move, unable to call out, Flo put a hand to her side and stared, as if in disbelief, at the blood. Then, in slow motion, she fell – down, down, down . . . Flint, like a limpet, still clung to her dying body and they hit the ground with a sickening thud together.

As the grinning human mask approached, white teeth gleaming in ebony face, Flo gave one final convulsive heave and then was still. Screaming, fighting, biting, Flint was pushed into the dank, evil-smelling sack, down, down, down. Yet even in the darkness I could see the black shadow of the man . . .

I woke in a cold sweat with the blanket over my face. So vivid was the nightmare that I could not sleep again. Yes – a nightmare. But it does happen, again and again, in parts of West and Central Africa. In many areas chimpanzee flesh is a much prized delicacy and there are horrifying tales of infant chimps tied up beside the sliced-up bodies of their mothers in the meat market, on sale for fattening and future consumption by the protein-starved Africans. Also chimpanzee infants are in great demand by the medical research laboratories of Europe and the United States and, God forgive us, this is the way we capture them too – by shooting their mothers. How many mothers must creep away through the dense upper canopy of the forest, mortally wounded, and die later so that their doomed infants are orphaned? How many babies that survive both the shot and the crash to the ground must die within the first few days of capture, shattered and desolate? I estimate that for every live infant that arrives in the Western world, an average of six others have lost their lives.

There is another shadow, too, which is spreading over the chimpanzee to-day, for, with the spread of agriculture and forestry, the habitat as well as the life of the chimpanzee is threatened. Forests are cleared to make way for cultivation and food-trees poisoned to leave more space for better timber trees. Moreover, since chimps are susceptible to all the infectious diseases of man, wherever their populations are near new human settlements the apes are endangered by epidemics.

Fortunately some people are waking to the dangers threatening chimpanzees in the wild. The enlightened governments of Uganda and Tanzania offer protection to their chimpanzee populations, and a recent international conservation meeting agreed to put these apes on to their list of endangered species, needing protection. Some programmes are being set up in order that chimpanzees needed for research can be bred in captivity and, if large and successful colonies can be established for this purpose, it will do much to relieve the constant drain on wild populations. The chimpanzee is, indeed, only one of the many species threatened with extinction in the wild; but he is, after all, our closest living relative and it would be tragic if, when our grandchildren are grown, the chimpanzee exists only in the zoo and the laboratory. A frightening thought; since, for the most part, the chimpanzee in captivity is very different from the magnificent creature we know so well in the wild.

Many zoos to-day are beginning to display their chimpanzees in groups in fairly spacious enclosures, but there are still many apes imprisoned in the old-fashioned concrete and barred cells. One summer I got to know two old zoo chimpanzees, a male and a female. They were housed in a very small cage with an indoor and an outdoor section, separated by a steel door. It was very hot that summer but, though they were kept outside with the dividing door closed, there was no awning and when the sun was overhead the concrete burnt to the touch. The chimps had no branch to sit on, only a very small wooden shelf which held one of them at a time – of course it was always the male who sat on this shelf when the sun was hot. They were fed only twice a day – once in the early morning and once in the late afternoon. Their water supply was normally finished by ten o'clock in the morning but it was not replenished.

How long ago, I thought as I watched them, they must have forgotten the vines, the soft ground, the swinging in the branches, and the excitement of rushing through the forest clinging to their mothers' bodies. By now their only pleasure probably lay in their food, but they could not enjoy the fishing for juicy insects, or the flavour of fresh-caught meat; and never again could they climb, with grunts of pleasure, to eat their fill of sun-ripened fruit in a cool forest tree. Also, how long the gap between meals must be for animals who, by nature, like to eat on and off throughout the day. Except for feeding they had nothing to do save listlessly groom each other or themselves. They could not even escape from each other, these two, not even for one minute; the male could never relax in the comfort of male companionship, and the female could never get away from male society.

\* \* \*

A chimpanzee in such a zoo is probably not too dissimilar to a human who has been in prison for many years and has no hope of release. Even in a better zoo, where the chimpanzee may be part of a slightly larger group and have a bigger concrete enclosure, he is a very different creature from the chimpanzee we know at the Gombe Stream. The zoo chimp has none of the calm dignity, the serenity of gaze or the purposeful individuality of his wild counterpart. Typically, he develops odd stereotypes in his behaviour – as he walks he may give one hand a slight rotation to the side, always the same hand, always to the same side. After shuffling across the confined space of his cage he may hit the iron frame of his door, always in the same place in the same way with the same rhythm. It is a pitiful vestige of the magnificent and impressive charging display of the wild male chimpanzee.

Most people are only familiar with the zoo or the laboratory chimpanzee. This means that even those who work closely with chimpanzees, such as zoo-keepers or research scientists, can have no concept or appreciation of what a chimpanzee *really* is. Which is, perhaps, why so many scientific laboratories maintain chimpanzees in conditions which are appalling, housed singly for the most part in small concrete cells with nothing to do day in and day out except to await some new – and often terrifying or painful – experiment.

I should make it clear that I am not trying to say that we should never use the chimpanzee as an experimental animal. Recent work by physiologists and bio-chemists has shown that biologically – in the number and form of chromosomes, blood protein, D.N.A. and so on – the chimpanzee is very close indeed to man. He is, indeed, as closely related to us in some respects as he is to the gorilla. Because of this, he is probably the only really effective substitute when, for ethical reasons, research cannot be carried out on humans. Kuru, a strange trembling illness of New Guinea, was one of the mysteries of the medical world and claimed countless victims. Research with chimpanzees established it as a slow-acting virus disease and made the present dramatic cure possible. Also, because of the similarity of the chimpanzee brain to the human brain, experiments carried out with chimpanzees will undoubtedly be of great value to those scientists who are grappling with the problems of human mental illness. Some people may not be aware either of the prevalence of or the tragedy involved in disorders of the brain such as, for example, schizophrenia or severe depression. It is not only the victims themselves who are affected but also their families and friends. Chimpanzees may serve as experimental models for some of these disorders and this assists science in its fight to alleviate human suffering.

On the other hand, whilst I agree that the chimpanzee should take part, if absolutely necessary, in some such experiments, I also feel strongly that laboratory chimpanzees should be given much-improved living conditions. Surely if we want to use this ape as a guinea pig in medical research, whether it be in connection with kidney transplants, drug addiction or long-term effects of the Pill; if we desire him to help man in his conquest of space – then we should make every effort to see that he is a well-treated and honoured guest in our laboratories. His living space should be large, he should be provided with equipment to alleviate his boredom, tempting and tasty food and, when possible, he should be housed with companions. Sometimes Hugo and I feel that the only way in which we might effect an improvement in the condition of laboratory chimpanzees is to take those responsible for their upkeep to see our chimpanzees at the Gombe Stream.

*Chapter 21* Family Postscript

Eventually the detailed understanding of chimpanzee behaviour that will result from our long-term research at the Gombe will help man in his attempts to understand more of himself. Hugo and I are convinced of this. Yet it is not only for this reason that we continue our work, year in, year out; we are also fascinated by the chimpanzees as individual beings. We want to know how Fifi, whom we first knew as a small infant, will look after her own children; whether Flo survives to be a grandmother and, if so, how she will react to Fifi's infant; what happens to Flint when his mother finally dies; whether Figan, one day, may become top-ranking male. In fact, we want to continue observing the chimpanzees for similar reasons to those which impel one to read on to the end of an exciting novel.

These days we cannot spend too much time at the Gombe because we have a child of our own and, as I mentioned earlier, chimpanzees have been known to prey on small human children. When our own son, little Hugo – better known as Grublin or Grub – was very small, we kept him inside a building in a cage up at the observation area. When Rodolf and Humphrey and young Evered looked in through the windows and rattled the bars, with their hair on end and their mouths tight-lipped and ferocious, we knew that if they had had the chance they would have seized Grub. We could not blame them for this; in captivity, where chimpanzees grow up in human rather than ape society, they are as tolerant and delighted with human as with chimpanzee babies. But at the Gombe Stream, the white-skinned apes that the chimpanzees have learned to tolerate – even to trust – are not normally associated with infants. Rodolf did not see Grub as my precious baby – merely as a tempting meal.

When Grub got older we could no longer keep him in his babyhood cage, so we built a much larger one down on the beach where the chimps seldom roam. This 'cage' is attached to a prefab house, thatched with grass, and it is cool and airy. When Hugo or I are with him then Grub can run on the white sandy beach and splash in the sparkling lake, but if we are both up with the chimpanzees then Grub, together with the two Africans who look after him, stays in the safety of the cage. Grub is happy at Gombe – he repeatedly tells everybody that it is his 'favourite place in all the world'. This is lucky since in the immediate future Hugo and I plan to return there on a more permanent basis.

For the past few years Hugo has been working on a study of the large African carnivores: just as Hugo has always involved himself in the chimpanzee research, so for a while I became involved in his project. From the time when we married, Hugo and I vowed to be and to work together as much as we possibly could for, to our way of thinking, two people with a harmony of interests and similar goals in life are the most likely to enjoy to the full the richness which marriage can offer. So, for a while, I watched hyenas, and that experience has helped to broaden my outlook, has brought me back to contemplate chimpanzees with a fresh eye, and been of value to my interpretations of chimpanzee behaviour.

Despite the fact that I have been away from Gombe for months at a stretch there has, of course, been no break in the records – thanks to our students. It is never quite the same to hear about something as actually to witness it, but it is still exciting, and most people who work at the Gombe Stream share our enthusiasm for the chimpanzees as individuals, thus their accounts are vivid as well as accurate. Also I have been lucky during my periodic visits. The chimps have seemed to stage exciting events almost as though they knew I was there. So each month brings in another instalment, another chapter of the chimpanzee saga. Before I end I want to gather up some of the threads concerning those chimpanzee characters whom I have chosen to cover in this book.

Let me first describe what has happened to Olly and her family. Almost exactly a year after losing her small infant during the polio epidemic, old Olly had yet another baby, but it was two months' premature and still-born so that, once again, Gilka was deprived of a younger sibling. Six months after that Olly disappeared and, eventually, we were forced to believe her dead. Gilka, still only seven years old, spent much time wandering around by herself.

Soon after Olly's death, Gilka developed a swelling on her nose. At first it seemed to be painful, for when another youngster approached to play Gilka often pulled back, her eyes screwed shut as though in anticipation of being hurt. As the weeks went by it seemed to stop hurting her, but it got larger and larger. A year later Gilka's whole face was hideously deformed, her nose swollen into a gigantic protuberance. Small bumps appeared on either side of it, and on her brow. Poor Gilka, who had been one of the most pretty of the youngsters with her long silky hair, pale heart-shaped face and small pointed white beard, looked like some nightmarish, grotesque gnome.

We showed photos of the condition to a variety of medical friends and finally, after much discussion, resolved to try at least to find out the nature of the growth, even though we feared there would be little we could do. This is not the place to describe the methods we employed first to tranquillise and then to anaesthetise Gilka. Suffice it to say that, with the help of the well-known veterinarians, Sue and Tony Harthoorn, Professor Roy and Don Nelson, we found that Gilka was afflicted with a fungus disease which, at the present time, we are treating with antibiotics. Gilka, despite the operation, has lost no confidence in humans, and our students have no difficulty in following her around in the mountains.

We all hope desperately that Gilka will recover. For one thing we are very fond of her. For another, she and Evered represent one of the few sibling pairs whose relationship we have been able to study in detail over a long period. For some three years before Olly's death there had been little contact between her two children, but during the years following the loss of their mother Evered and his young sister seem to have grown closer. Quite often the two wander around together. Moreover, they spend a good deal of time grooming each other whereas the other young males, Evered's contemporaries, very seldom groom Gilka; when they do it is normally a brief session.

Whether or not the friendship between Gilka and Evered will persist we cannot guess. Pepe and Miff, Marina's offspring, also moved around together frequently during the year following their mother's death, but gradually they seemed to drift apart. Unfortunately Pepe died a year after his little sibling Merlin, leaving Miff, the sole survivor of Marina's family. But Miff, when she was between ten and eleven years old, had a daughter. We were amazed at the efficient way in which she handled her infant. Each of the other new mothers we have watched has been somewhat awkward and uncertain, at least for the first few days after her infant's birth. Melissa, indeed, had seemed quite dazed. Miff, however, accepted her baby as a matter of course; she paid her about the same amount of attention as old and experienced Flo devoted to her last infant. We cannot help but wonder whether, perhaps, Miff's experience with her small orphaned brother gave her this extra self-confidence.

For several years, as mentioned in previous chapters, we have wondered whether, perhaps, a sibling relationship lay behind some of the close friendships between pairs of adult males. We were, therefore, particularly anxious to follow the development of the relationship between Flo's elder sons, Faben and Figan. When they were both adolescent they quite often played together, though Faben tended to be rather rough with his younger sibling. When Faben became a socially mature male, however, about thirteen years old, he stopped interacting with Figan almost entirely. Hugo and I began to question whether our theory concerning friendships between males might, after all, be wrong.

Faben has a magnificent physique and during his first year as a mature male his charging displays, often performed in an upright posture, were splendid. Then he was stricken by polio and completely lost the use of his right arm and hand. Up to that time Figan, as the younger brother, had always been dominated by Faben, but it was typical of Figan's particular brand of intelligence that he took almost instant advantage of Faben's disability to better his own status. For the first few days after Faben appeared with his useless arm, Figan seemed to pay him no particular attention. But then he began deliberately to harass his elder brother.

The first time I witnessed this, Faben was sitting peacefully in a tree grooming himself. Figan stared towards him, slowly walked over and climbed the tree. Then he commenced a wild display, shaking the branches and rocking the whole tree so violently that Faben, who had at first ignored his young brother's behaviour, began to scream. Faben had not had a chance at that time to adjust to life with one paralysed arm: within a few moments he was shaken right out of the tree. Twice more during the next few days

Figan displayed at Faben in the same sort of way. After this the elder brother's attitude underwent a remarkable change: the next time I saw the brothers meet, Faben, as he saw Figan approaching, hurried towards him with submissive pant-grunts. Figan just sat there and, as his elder brother crouched before him, patted him reassuringly on his head!

This state of affairs lasted for a while but gradually Faben became better adjusted to his handicapped state. He had been able to walk completely upright from the start, thus preventing his paralysed arm from trailing on the ground; and gradually he became even more expert in this man-like method of locomotion. He was soon able to keep up with groups of adult males for quite long distances, and he began to regain something of his former agility when swinging through the trees. Also his charging displays gradually regained their former magnificence. It seemed that Figan must have noticed this physical improvement in his brother for he kept well out of Faben's way for a while. Two years later there was no sign, when the two interacted, that Figan had ever been able to dominate his older brother.

It was when Figan himself was promoted to the rank of a socially mature young male that, for the first time since their adolescence, we noticed that the relationship between the two brothers was becoming friendly. They were quite frequently observed travelling about together and lengthy sessions of social grooming between the two became commonplace. Perhaps, we thought, we had been right in our theorising after all: perhaps this was the start of a friendship such as that between David Greybeard and Goliath, Mike and J.B., Mr McGregor and Humphrey; a friendship which would persist, which would ensure one brother running to help the other in times of trouble. We could only wait and watch.

During the next year the relationship between Figan and Evered (who was, it will be remembered, about a year older) became increasingly tense; these two young males often directed displays towards each other. These conflicts, however, were not seen to result in physical combat until one never to be forgotten day. The incident started when Evered arrived, his hair bristling, to find Figan and Faben in camp together. When Evered advanced towards Figan the slightly younger male hurried up to his brother. Faben embraced him and then turned back towards Evered. A moment later the two brothers, united, charged towards Evered uttering fierce *waa* barks and chased him right out of the camp clearing. Screaming, Evered fled through the forest and took refuge up a tall tree. The brothers returned to camp.

For the next few minutes first Faben and then Figan kept displaying across the clearing uttering loud pant-hoots, dragging and flailing branches and drumming. Once there were a few rather subdued hoots from Evered in his tree.

About half an hour after leaving camp, Evered moved back towards the clearing, but Faben at once charged towards him and displayed, Figan hurried to join his brother, and the two of them raced flat out after Evered who quickly retreated up his tree. This time the brothers followed him and sat, their hair bristling, about five yards from their victim. All three were silent, but Evered's lips were drawn way back in a grin of fear.

Presently Figan advanced a few steps towards Evered who, with soft grunts, moved a few yards along his branch and then sat again. A few moments later, and still in silence, Faben moved towards Evered who squeaked nervously, moved farther out along his branch, and then sat. Presently Figan also advanced on Evered, who gave low grunts but could retreat no farther. The brothers sat facing him, silent and still.

All at once, about five minutes after the three had climbed into the tree, the battle erupted, as Faben and Figan together leapt, with loud *waa* barks, towards Evered. With screams, Evered hurled himself into a neighbouring tree, the brothers following close behind. There was a short chase and then Figan grabbed on to Evered at the extreme end of a thin bendy branch and, as the two grappled, Faben joined in the fight as well.

It was an incredible battle to watch as the three young males, in a screaming tangle, somehow managed to leap from one branch to another. Meanwhile, Flint had climbed into the tree and, keeping at a safe distance from the fight, stamped along the branches and uttered high-pitched childish *waa* barks whilst, down below, old Flo displayed on the ground, stamping and shaking vines and barking in her hoarse and ancient voice.

Suddenly Figan and Evered, still fighting, fell together. It seemed that they almost drifted through the air to crash into a dense tangle of vines some thirty feet below. Faben instantly swung and leapt down after them, and Evered, screaming loudly, fled away through the forest. Figan and Faben, followed by Flo and Flint, pursued him for some way and then, one after the other, displayed, uttering loud pant-hoots and drumming. A couple of hundred yards away, down near the stream, Evered was still screaming and whimpering.

During that encounter Figan hurt his arm, scraped some skin off his knuckles and lost great patches of hair. Evered, however, came off much the worst; he split open one side of his face, from the corner of his lip right up his cheek. Everyone feared that he would remain with a permament disfigurement, a sort of hideous grin that would leave his face as ghastly to look at as that of his sister Gilka. To our surprise, however, it healed up with scarcely a mark to show.

Meanwhile our research team was gathering fascinating new information on the subject of male dominance. It was not so long after old McGregor's death that his presumed young brother Humphrey developed into a truly massive male, and the larger he grew the more aggressive he became. By 1968 Humphrey had all the females and adolescent males very much in awe of him. Often, when Humphrey and Mike were together in a group, newcomers would hurry to pay their respects to Humphrey before greeting Mike himself. During the following year, Humphrey gradually became dominant even over old Rodolf and Leakey and Goliath. But still he himself respected Mike as much as ever.

Figan was particularly in awe of Humphrey, but, at the same time, he began to show a good deal less respect for Mike. Mike began to get very worried about this young upstart. When Mike displayed, the other chimps still rushed out of his way – all but Figan who sat quite calmly, his back to Mike. This happened again and again, and Mike became increasingly uneasy. When Figan was near he displayed more and more frequently, and most of his displays were directed towards the young male. Indeed, he swayed the very branches on which Figan sat, his back still firmly turned towards Mike. Yet Mike, it seemed, dared not actually attack Figan.

There was one memorable occasion when Figan ventured to approach a pink female of whom Mike was being strangely possessive. When Mike began threatening Figan, shaking the branches of the tree, Figan actually displayed back and, in the resultant confusion, both males rushed screaming from the tree. Then Mike, with a grin of fear, actually hastened to embrace Humphrey, seeking reassurance from this contact with the other male.

Shortly after this some incident occurred, unseen by human eyes, which put Figan firmly back in his place. After this Figan once again rushed out of the way when Mike displayed and hastened to greet him with submissive gestures. No sooner had Figan been subdued, however, than Evered, a year older than Figan, began to challenge Mike's supremacy. Just as Figan had done before him, Evered sat and ignored Mike's displays. And just as Mike had been worried by Figan's insubordination, so it was with Evered.

Mike to-day is still the dominant male, but he is very uneasy when he is on his own with Figan or Evered. Only when one of the old-timers is with him, such as Rodolf or Leakey, does Mike seem at all relaxed in the presence of these two young males. It is a strange situation; if either of the two did engage in battle with Mike, and dominate him, the old male would no longer occupy the top-ranking position. But neither would the youngster who had usurped his position – for both Figan and Evered still show tremendous respect for Humphrey. Soon then we may have a situation where no single male is dominant in all situations – certainly something is going to happen very soon indeed.

Hugo and I suspect that eventually Figan will become the top-ranking male – possibly after Humphrey has had an innings. For Figan is not only more intelligent than Evered – he also has the backing of a large family. The proximity of Faben will probably always give him that feeling of confidence that David Greybeard once gave to Goliath.

Goliath himself is a somewhat tragic figure these days. For four years after Mike had usurped his top-ranking position Goliath remained a very dominant male, but then he became ill and lost much weight, added to which his old friend David Greybeard died during an epidemic of flu. To-day Goliath is dominated not only by all the adult males, but by most of the adolescents too. He spends much time wandering about alone, or with old Rodolf or Leakey. In the past two years alone his weight has decreased from one hundred to a mere seventy pounds.

Soon, I suspect, Goliath will be no more. Nor can Flo, amazing though she is, be expected to survive much longer. We know so much about these chimpanzees, we have shared so much of their lives, that it is always sad when they die. For me, of course, the saddest loss was when David Greybeard died. For David was the first chimpanzee to accept my presence and permit me to approach him closely. Not only did he provide me with my early observations of meat-eating and tool-using and thus help to ensure further funds were available for my work, but he was the first to visit my camp, to take a banana from my hand, to permit a human hand to touch him. I have mentioned earlier the mistake Hugo and I made in allowing Flint to touch us, in encouraging Fifi and Figan to play with us. We endangered not only the validity of subsequent research but, worse, the safety of those students who would follow in our footsteps at the Gombe Stream. To this day Flint and Figan sometimes try to initiate play with their human observers.

But I do not regret my early contact with David Greybeard; David, with his gentle disposition, who permitted a strange white ape to touch him. To me it represented a triumph of the sort of relationship which man can establish with a wild creature. Indeed, when I was with David I sometimes felt that our relationship came closer to friendship than I would have thought possible with a completely free wild creature, a creature who had never known captivity.

In those early days I spent many days alone with David. Hour after hour I followed him through the forests, sitting and watching him whilst he fed or rested, struggling to keep up when he moved through a tangle of vines. Sometimes, I am sure, he waited for me – just as he would wait for Goliath or William. For when I emerged, panting and torn from a mass of thorny undergrowth, I often found him sitting, looking back in my direction. When I had emerged, then he got up and plodded on again.

One day, as I sat near him at the bank of a tiny trickle of crystal-clear water, I saw a ripe red palm nut lying on the ground. I picked it up and held it out to him on my open palm. He turned his head away. But when I moved my hand a little closer he looked at it, and then at me, and then he took the fruit and, at the same time, he held my hand firmly and gently with his own. As I sat, motionless, he released my hand, looked down at the nut, and dropped it to the ground.

At that moment there was no need of any scientific knowledge

Flo and Fifi startled
by a sudden noise.
Physical contact is
reassuring, even when
only between two hands

Submissively Melissa kisses J.B.'s reassuring hand

Presenting is a sign of submission

Two chimpanzees embrace in greeting

Rodolf attacks a female, then immediately reassures her with an embrace

Gilka, eighteen months after contracting a fungus disease

Flo and Flame

to understand his communication of reassurance. The soft pressure of his fingers spoke to me not through my intellect but through a more primitive emotional channel: the barrier of untold centuries which has grown up during the separate evolution of man and chimpanzee was, for those few seconds, broken down.

It was a reward far beyond my greatest hopes.

# *Appendix 1* Facial Expressions and Calls

The DISPLAY FACE is shown by aggressive chimpanzees, especially during their charging displays or when attacking others. It is not accompanied by calling.

The PLAY FACE is shown by chimpanzees during play. When a game becomes vigorous the upper lip is often drawn back and up so that the front upper teeth are also exposed. The play face is frequently accompanied by a series of grunting sounds or LAUGHING.

Two of the facial expressions typically shown by chimpanzees as they utter a series of PANT-HOOTS. Pant-hoots are a series of *hooo* sounds connected by audible intakes of breath, gradually getting louder and usually ending with *waaa* sounds also connected by panting intakes of breath. Picture (a) shows the expression during a *hooo* and (b) during a *waaa*. Pant-hoots are given in a variety of contexts, especially when chimpanzees arrive at a food source, join another group or cross from one valley to another. Pant-hoots serve as a contact call between spread-out individuals or groups: sometimes chimpanzees peacefully feeding in a tree will give this call at which another group, some way away, may respond. Quite often when groups of chimpanzees are sleeping within earshot of each other they may exchange pant-hoots during the night, particularly when there is a bright moon.

(a)

(b)

GRINNING. (*Right above*) a FULL OPEN GRIN (upper and lower front teeth showing, jaws open) is usually shown by a chimpanzee who is frightened or very excited: during and after an attack, when a high-ranking male displays close to a subordinate, when a group of chimpanzees is confronted with a huge pile of bananas and so on. If the chimpanzee is less frightened or excited the upper lip may be slightly relaxed to cover the upper front teeth —this is the LOW OPEN GRIN. The full and the low open grins may alternate rapidly with one another and are nearly always accompanied by loud SCREAMING.

(*Below*) A FULL CLOSED GRIN (upper and lower front teeth showing, but with jaws *closed*) is the expression of a chimpanzee who is probably less frightened or excited than one showing either of the *open* grins. A full closed grin may alternate with the low *open* grin or with a LOW CLOSED GRIN in which only the lower front teeth show. *Closed* grins are usually accompanied by high-pitched SQUEAKING sounds which readily change to screams or WHIMPERS. Sometimes, however, a low ranking chimpanzee may approach a superior in silence while showing a closed grin. If the human nervous or social smile has its equivalent expression in the chimpanzee it is, without doubt, the closed grin.

POUTING. As screams or squeaks change to WHIMPERING—a series of *oo* sounds repeated rapidly at different notes of the scale—the chimpanzee shows this HORIZONTAL POUT. After being attacked a youngster may gradually stop screaming and squeaking and start to whimper. Whimpering, however, does not only follow screaming and squeaking. A soft single syllable HOO WHIMPER is the contact call between a mother and her infant. It is the sound made by a chimpanzee who desires food from a superior, who wants his companion to groom him, or who is frustrated in other ways. This soft hoo whimper is accompanied by a POUT when both lips are pursed together and pushed forward. If the frustration is not relieved the call is repeated more frequently and more loudly until it becomes normal whimpering and the lips become increasingly parted. A POUT looks almost identical with the expression shown in Picture (a).

*Other expressions and calls.*

Chimpanzees GRUNT in a variety of contexts; these grunts may be high-pitched or low-pitched. Usually there is little change in facial expression during grunting, though the lips and jaws may be slightly parted. Grunts are given during feeding (FOOD GRUNTS), grooming, and as close-range contact calls between the individuals of a peaceful group.

A series of rapid grunts, connected by audible intakes of breath, are known as PANT-GRUNTS. A subordinate chimpanzee is likely to pant-grunt as he approaches a superior during a greeting or after being threatened or attacked. During pant-grunting the jaws are often partly open, the teeth normally concealed by the lips. If the superior behaves at all aggressively during the interaction pant-grunting quickly becomes squeaking or screaming and the chimpanzee grins.

Loud BARKING often occurs when a group is socially excited: while some individuals pant-hoot others may bark. Very loud FOOD BARKS often occur as chimpanzees arrive at a favoured food source and during the first few minutes of intensive feeding. The jaws open slightly as each bark is emitted, and the lower front teeth may show slightly.

When mildly threatening another chimpanzee (or animal of another species including humans) a chimpanzee utters a SOFT BARK. This sounds very like a single quiet cough; the jaws are very slightly parted with the lips covering the teeth.

When threatening more vigorously the chimpanzee often utters a loud WAA BARK—the facial expression is very similar to that shown in picture (b).

The WRAAAA CALL of the chimpanzee is, to me, one of the most savage sounds of the African jungle. It is a long drawn out clear sound, pitched rather high and is made when chimpanzees come across something unusual

**245**

or slightly disturbing in the forest. It was with this call that the chimpanzees acknowledged my approach in the early days once they had got over their initial terror of me. They may call thus when they come across a herd of buffalo or a dead chimpanzee. The facial expression is, again, very similar to that depicted in (b). In similar contexts some chimpanzees may utter long, loud screams. Audible PANTING, a little similar to laughing though without the grunting quality, is sometimes heard when chimpanzees are grooming each other intensively. Some males utter quite loud COPULATION PANTS during mating. No clear-cut facial expression accompanies panting.

## *Appendix 2* Diet

Like man, the chimpanzee is an omnivore and feeds on vegetables, insects and meat.

*Vegetable Foods*
Over 90 different species of tree and plant used by the Gombe Stream chimpanzees for food have already been identified. They have been seen eating over 50 types of FRUIT and over 30 types of LEAF and LEAF BUD. They also eat some BLOSSOMS, SEEDS, BARKS and PITHS. Sometimes they lick RESIN from tree trunks or chew on wadges of DEAD WOOD FIBRE.

*Insect Foods*
Throughout the year the following kinds of insects may be eaten in large quantities:
3 species of ANT (*Oecophylla* spp., *Anomma* spp., *Crematogaster* spp.);
2 species of TERMITE (*Macrotermes* spp., *Pseudacanthotermes* spp.);
1 species of CATERPILLAR of a moth as yet unidentified.
These chimps also eat a variety of GRUBS—the larvae of different beetles, wasps, gall flies, etc.). BEE LARVAE are eaten when chimpanzees raid bees' nests and feed on HONEY.

*Birds' Eggs and Fledgelings*
Occasionally the chimpanzees take eggs or fledgelings from the nests of a wide variety of birds.

*Meat*
The Gombe Stream chimpanzees are efficient hunters: a group of about 40 individuals may catch over 20 different prey animals during one year. Most common prey animals are the young of BUSHBUCKS (*Tragelaphus scriptus*),

BUSHPIGS (*Potamochoerus porcus*) and BABOONS (*Papio anubis*), and young or adult COLOBUS MONKEYS (*Colobus badius*). Occasionally the chimpanzees may catch a REDTAIL MONKEY (*Cercopithecus ascanius*) or a BLUE MONKEY (*Cercopithecus mitis*).

Bushbuck

Bushpig

Colobus monkev

Baboon

*Minerals*
The chimpanzees sometimes eat small quantities of soil containing some salt (sodium chloride).

# *Appendix 3* Weapon and Tool Use

The chimpanzee uses the objects of his environment as tools to a greater extent than any other living animal with the exception of man himself. The drawings show:

1. Stick used as weapon.
2. Aimed throwing.
3. Investigation probe—chimpanzee sniffs end of stick after poking it into a hole in dead wood. If insect larva is detected the wood will

be broken open and the grub consumed. (A probe of this sort may also be used to investigate an unusual object such as a dead python).

4. Stick used to feed on safari ants. These ants have a very, very painful bite and the chimpanzee tries to prevent them crawling over his body whilst he plunges his stick into the ants' underground nest. A stick is also used to feed on ants living in hard football-sized nests constructed around the branches of trees.

5. Grass stems used to "fish" for termites.

6. Tool-*making*—leaves are stripped from a stem to make a tool suitable for termite fishing. In addition the edges of a wide blade of grass may be stripped off in order to make an appropriate tool.

7. Leaves, which the chimpanzee has made more absorbent by chewing, are used as a "sponge" to sop up rainwater that cannot be

reached with the lips. The initial modification of the handful of leaves provides another example of primitive tool-*making*.

8. Leaf sponge used to wipe remnants of brain from inside the skull of a baboon.

9. Leaves used to dab at bleeding wound on bottom. Sometimes, when a chimpanzee has diarrhoea, he uses leaves as toilet paper: he may also use them to wipe off mud, sticky foods, etc.

In addition to the above examples of objects used by the Gombe Stream chimpanzees as tools, the chimpanzees were observed to use stout sticks as "levers"—to enlarge the opening of an underground bees' nest and frequently, to try and pry open banana boxes at the observation area. One chimpanzee used a twig as a toothpick and one picked its nose with a piece of straw.

# *Appendix 4* Some Milestones in the development of Chimpanzees

Some chimpanzees, like some humans, develop faster than others: moreover, the behaviour of a chimpanzee's mother undoubtedly has a marked effect on his physical and social development; the same holds good for human mothers too. A mother, for instance, may be permissive or restrictive towards her child's early attempts to walk; she may be tolerant or nervous of his initial contacts with others of his kind. The ages given here for the appearance of the different physical or social developments in chimpanzees are the *earliest* at which they were observed in any of our youngsters.

(Month 2 refers to the second month of life: i.e. the infant is between 4 and 8 weeks of age: similarly year three refers to youngsters between 24 and 36 months of age).

| | Month first seen | | Month first seen |
|---|---|---|---|
| Sucks thumb | 2 | Kisses another | 5 |
| Mother may tickle occasionally and briefly | 2 | Attempts to groom another, inefficiently | 7 |
| Stares at object; reaches towards it | 2 | Attempts to make a nest | 8 |
| Struggles to pull from mother | 2 | Attacked mildly by another | 8 |
| Stands upright holding on to mother | 2 | Mounts and thrusts "pink" female | 8 |
| Pushes and pulls itself forward on mother's body | 3 | Runs at and hits another infant aggressively | 16 |
| Reaches towards object and grasps, showing co-ordination | 3 | Reassures another in correct context | 16 |
| First tooth | 3 | Grooms with adult technique | 18 |
| Mother plays frequently and for longer at a time | 3 | | *Year* |
| | | Charging display and "rain dance" in correct context | 3 (early) |
| Infant shows play face and laughing during tickling | 3 | Violent attack on another youngster | 3 (early) |
| Chews and swallows first piece of solid food | 4 | Attempts tool-using in correct context | 3 (early) |
| Reaches to play with mother's hand during game | 5 | Weaned | By end of 5 |
| Starts to ride on mother's back* | 5 | Starts to lose milk teeth | 6 |
| Takes first step | 5 | May start to move around for short periods without mother | 6 |
| Mother-infant contact broken | 5 | Attains puberty | About 8 or 9 |
| Climbs up sapling or branch | 5 | First infant born | About 11 or 12 |
| Kidnapped (a sibling may kidnap earlier, during 4th month) | 5 | Male becomes fully socially mature | About 15 |

*One infant (Pom) rode in the dorsal position at 8 weeks: this was most unusual and was because she hurt her foot. See p. 140.

# *Appendix 5* In what ways can the study of chimpanzees at Gombe benefit mankind?

1. The chimpanzee is our closest living relative: recent research suggests that man and chimpanzee may have had a common ancestor in the distant past. Biochemists have shown that in some ways the chimpanzee is closer to man than he is to the gorilla. In particular the circuitry of the chimpanzee brain is remarkably like that of man. And observations at Gombe have highlighted striking similarities in the behaviour of chimpanzee and man, notably in non-verbal communication patterns. Thus a thorough understanding of chimpanzee behaviour will be of value in our efforts to understand our own.
2. The problem of human aggression is of vital importance. Before we can hope to control violence we must understand it. The studies of chimpanzee aggression in progress at Gombe may prove of immense significance.

3. The methods by which we raise our children, and the welfare of orphans and socially deprived youngsters, are also problems of major concern. Our studies at Gombe, both on different mothering techniques and the behaviour of abnormal youngsters, have already proved of interest to child psychologists and psychiatrists. Why do some adult chimpanzees maintain closer bonds with their families than others? The answer may be significant to the understanding of human family problems.

4. Human adolescence is a difficult time, as it is for chimpanzees. If we can learn more about the psychological and physiological changes in chimpanzees this may help us to understand and help our own teenagers.

5. Human mental illness causes untold suffering—and is on the increase. Scientists working on, for example, the prevention and cure of clinical depressions, hope to induce the similar symptoms in the chimpanzee and thus use him as an experimental model. However, such a scientist, before he can hope to measure in full the success or failure of a given treatment, must have knowledge of the behaviour of a "normal" chimpanzee. In addition he must be aware of the conditions which a chimpanzee, in captivity, must enjoy if it is to show normal behaviour. We hope our work at Gombe will provide this information.

6. Professor Hamburg, in collaboration with us, is setting up a large outdoor chimpanzee enclosure at Stanford. Initially we shall continue and elaborate some of the Gombe research projects on aggression, adolescent problems and maternal behaviour. Ultimately Professor Hamburg will work also on human mental illnesses. This will bring our work at Gombe closer to my ultimate goal—using our knowledge of chimpanzee behaviour to benefit humanity.

7. Research at Gombe has encouraged collaboration between students from many different sciences—ecology, ethology, anthropology, psychology and psychiatry. The participation of students and faculty members from the Universities of Dar es Salaam, Cambridge, Stanford and Amsterdam will encourage further exchanges of ideas. Such interdisciplinary collaboration will undoubtedly give rise to new lines of enquiry into many problems of human behaviour.

8. A thorough knowledge of the behaviour and diet of wild chimpanzees will be of great importance to the successful maintenance of breeding colonies in captivity. Such colonies are vital if we are to stop the constant drain on natural resources caused by the demands of science.

9. Conservation of the fauna of the world is another pressing problem. In order adequately to protect animals it is necessary to understand their requirements. Ecological and behavioural studies at Gombe are yielding a wealth of information that will be valuable to those responsible for the management of other Parks and Reserves where there are chimpanzees.

10. We aim to establish an observation area for tourists in the Gombe National Park. Thanks to the courtesy of the Tanzanian Government, its President, Julius Nyerere, and the Tanzanian National Parks, visitors from all over the world will have a unique opportunity to observe chimpanzees in all the splendour of their freedom.

# Index

# Index

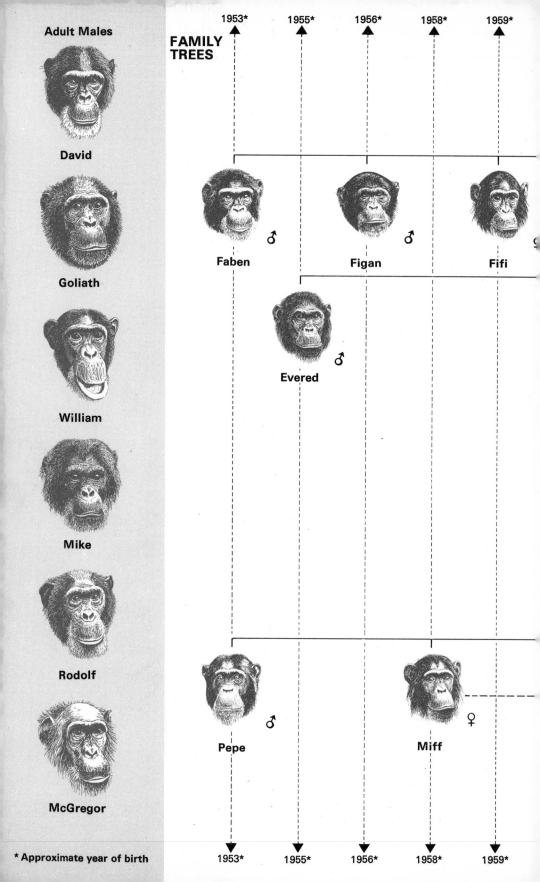

**Adult Males**

David

Goliath

William

Mike

Rodolf

McGregor

* Approximate year of birth

FAMILY
TREES

1953*    1955*    1956*    1958*    1959*

Faben ♂    Figan ♂    Fifi ♀

Evered ♂

Pepe ♂    Miff ♀

1953*    1955*    1956*    1958*    1959*